# Electrochemistry
## for Biomedical Researchers

# Electrochemistry
## for Biomedical Researchers

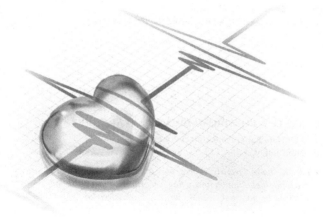

## Richie L. C. Chen
### National Taiwan University, Taiwan

臺大出版中心
NATIONAL TAIWAN UNIVERSITY PRESS

World Scientific

*Published by*

National Taiwan University Press
1, Sec 4, Roosevelt Road
Taipei, Taiwan

and

World Scientific Publishing Co. Pte. Ltd.
5 Toh Tuck Link, Singapore 596224
*USA office:* 27 Warren Street, Suite 401-402, Hackensack, NJ 07601
*UK office:* 57 Shelton Street, Covent Garden, London WC2H 9HE

**British Library Cataloguing-in-Publication Data**
A catalogue record for this book is available from the British Library.

ISBN 978-981-4407-99-1

Typeset by Stallion Press
Email: enquiries@stallionpress.com

Printed in Singapore.

# Preface

Have you ever noticed that from the simplest pH electrode to capillary electrophoresis used in the human genome project, almost all bio-related techniques are based on electrochemistry? What is more, even the recent progress in the battery industry has further elevated the importance of this knowledge domain.

However, for most students without a substantial chemical background, it will be a long way to go before fully understanding the underlying electrochemical principles. As in other branches of physical chemistry, most physical quantities used in an electrochemistry textbook are often dealt with very seriously, or even nervously. For example, concentration is replaced with activity to handle free energy more precisely, so Debye-Hückel theory has to be introduced first before entering the target subject. This will certainly prolong the learning.

Mathematics is the common language of scientists; rather than intentionally avoiding the necessary mathematical treatments, it is better to provide "less nervous" and "more effective" explanations in a concept-by-concept manner for the most common electrochemical techniques, especially for those frequently encountered in biomedical studies. Following each chapter, I used to present at least one demonstration experiment designed to elucidate possibly obscured concepts. These experiments were excluded from this book to limit its size, but the details (along with the original version written in Japanese) can be downloaded from the website of our department or my lab.

Additionally, to make the book more complete and self-comprehensive, I have also supplemented the text with some background information and optional applications in the Appendices.

Finally, before finishing, I wish to thank all the colleagues in our department and the students over the years for all of their helpful and timely responses. After that, I should apologize to all the readers for enduring my Taiwanish English and clumsy drawings.

Richie L. C. Chen

*February 28, 2008*

*Biosensor Group*
*Department of Bio-Industrial Mechatronics Engineering*
*National Taiwan University*
*E-mail: rlcchen@ntu.edu.tw*

# Brief Contents

# Contents

# Chapter 01

# Membrane Potential

## Bio and electricity

Several kinds of aquatic animals such as electric eels or rays are capable of generating electricity. On each side of the gills of a Japanese electric ray there is an electric organ which behaves as a battery with about 30 V output voltage. Using this amazing bioelectricity, these queer creatures can easily catch their victims or just keep away from their predators. In this chapter, we will learn where this electricity comes from.

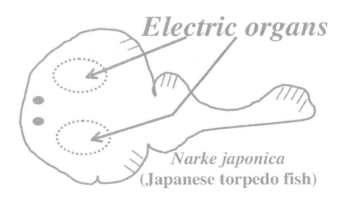

*Electric organs*

Narke japonica
(Japanese torpedo fish)

## A cell is, in itself, a concentration cell

"All living creatures live by eating negative entropy." As usually stated by some biophysicists, life is actually a thermodynamic process of sacrificing energy to maintain a low entropy state. A biomembrane, a bi-layered structure of phospholipids, is the barrier that suppresses thermal diffusion and keeps the "order" of molecules. Integrated onto the membrane are some proteins (ion pumps) which generate the difference in ion concentration across the membrane (active transport: from low to high concentration). There are also some proteins (ion channels) selectively permitting the diffusion across the ion barrier (passive transport: from high to low concentration) with high electrical resistance. Cells are therefore surrounded by these functional barriers from the start of their "independent" lives. The ionic compositions inside and outside the membrane are totally different.

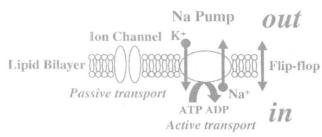

Resistance / area > 1 G$\Omega$cm$^{-2}$

Biomembrane : *Fluid-mosaic model*

It is known that connecting two compartments with differing ion concentrations will result in a concentration cell with a voltage difference. The voltage across a cell membrane can be measured with a micro-pipette filled with solution similar to that in the intracellular environment. The micro-pipette can be made by pulling a capillary at the glass transition temperature. After attachment to the cell surface, the attached "patch" of membrane can be extracted by mild suction.

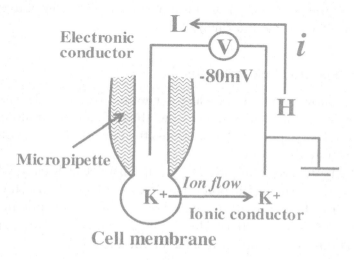

As measured with a voltmeter, the "intracellular potential" is about −80 mV relative to the outside environment. The 30 V output of the electric organ of a Japanese electric ray must result from the connection of quite a few (hundreds of) "concentration cells" in series.

## The dilemma of Kirchhoff's loop rule

The major ion channels on a cell membrane are potassium channels which selectively transport potassium ion through the membrane. Since the potassium in the intracellular compartment is usually higher than outside, there is a trend for potassium flux from the inside to the outside.

Based on Kirchhoff's loop rule, there must be an equivalent current flux from the ground side to the electrode inserted within the micro-pipette. But now, it is a flow of electrons (electronic current) not ions (ionic current). Electronic current will flow from high potential to low potential, so the potential inside should be lower than outside. That is exactly what the voltmeter shows.

However, if the measured potential is correct, the ionic current — the potassium flux — is now flowing from low potential to high potential. Is that reasonable?

### Battery: a pump of electric potential

The following simplified electric circuit gives us the key to solve the above dilemma. The current flows through a resistor along with a drop of electric potential, but, across the battery, the current flows from low to high potential.

Yes, a battery is just a "potential pump" in an electric circuit. In the experiment measuring the potential across a cell membrane, the ionic current flows from the intracellular compartment (low potential) to the outside (high potential). The cell membrane actually is a kind of potential pump, and therefore a battery in the circuit.

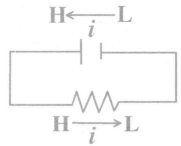

Potential Change with Current Flow

### From chemical energy to electric energy

The general rule of Mother Nature is to seek her own balance. In the intracellular space with a high potassium concentration, there is a trend for the efflux of potassium ion. But the trend is counteracted by the low electric potential generated by the concentration difference.

More precisely, the potassium ion inside the cell membrane is prone to "diffuse" following the concentration drop to the outside. At the same time, the concentration difference also generates an electric potential, and thus an electric field toward the inside. Therefore, a cation (potassium) tends to "migrate" from the outside to the inside. Finally, the thermodynamically favored behavior (diffusion) is balanced by the electrically favored behavior (migration).

To sum up, the difference in ion concentration (chemical energy) will spontaneously lead to a difference in electric potential (electric energy) to balance the thermodynamic trend. A flow of ionic current following the thermodynamic trend will result in converting chemical energy to electric energy. The "battery" is said to be discharging.

Now, the remaining task is to solve mathematically the efficiency of this energy conversion. Before that, let's have a glance at logarithms and free energy.

## The natural significance of logarithm

For survival, living creatures continuously receive a wide dynamic range of some environmental signals such as light and mechanical pressure. However, the resolving power for tiny differences in signal intensity will be sacrificed at the same time.

Taking vision as the example, human beings can hardly discriminate between the light intensity from a 20 W and a 40 W electric bulb, but they can see and work in either daytime or midnight. In reality, human beings do not directly sense light intensity (number of photons incident on a defined surface per unit time), but its logarithmic value.

Similar to vision, the "magnitude" of an earthquake is another example of the natural significance of logarithms. Human beings can feel according to the logarithmic value of a vibrational amplitude but not the amplitude itself.

Therefore, taking the logarithm is generally the first step in biological adaptation toward sensing diverse environmental signals with a wide intensity range in a more converged scale.

## Free energy: the thermodynamic trend for diffusion

Thermodynamically, there is a natural trend for diffusion from a high concentration to a low concentration to maximize system entropy. The trend is driven by thermal motion and therefore related to the absolute temperature. In the beginning of the 20th century, the American physicist Gibbs defined the free energy for one mole of freely diffusing molecules at a specific concentration as:

$$G = G^0 + RT \ln C$$

Where $G^0$ is the free energy at the standard state (1 bar, 25°C, 1 M); $R$ represents the gas constant; $T$ is the absolute temperature (°K); and $C$ is the concentration of the molecule. The energy is quite "biological" since it is linearly correlated with the logarithmic value of the signal, the concentration of the molecules. The molecules tend to diffuse from a location with higher free energy (more concentrated) to others with lower free energy.

## Calculating membrane potential

We are now standing on the point ready to calculate membrane potential ($V_m$) using the concept of free energy. Potassium ions are to diffuse from an intracellular compartment (intracellular potassium concentration, $K_i \approx 90\,\text{mM}$) to the outside (extracellular potassium concentration, $K_o \approx 3\,\text{mM}$); the free energy per mole transition will decrease as in the following:

$$\Delta G = RT[\ln(K_o) - \ln(K_i)] = RT \ln \frac{3\,\text{mM}}{90\,\text{mM}}$$

At the same time, potassium ions are transported from a location with low electric potential (intracellular potential, $V_i$) to one with high electric potential (extracellular potential, $V_o$); the increase in electric energy per mole will be:

$$\Delta E = nF(V_o - V_i) > 0$$

In which, $n$ is the charge carried by the ion; $F$ is Faraday's constant that is equivalent to the charge per mole of protons ($F \approx 10^5$ Coulombs).

Assuming energy conservation, the decrease in chemical energy ($\Delta G$) will be equal to the increased amount of electric energy ($-\Delta E$).

$$\Delta G = -\Delta E$$

$$RT \ln \frac{K_o}{K_i} = -nF \left( V_o - V_i \right)$$

$$V_m = V_i - V_o = \frac{RT}{nF} \ln \frac{K_o}{K_i}$$

$$= \frac{RT}{nF} \times \frac{1}{\log e} \times \log \frac{K_o}{K_i} \cong \frac{59\,\text{mV}}{+1} \times \log \frac{3}{90} = -87\,\text{mV}$$

Where $V_m$ is the membrane potential which is correlated with the logarithmic concentration ratio by a constant of 59 mV at 25°C. Cell membrane potentials may be raised by adding saline with a higher potassium level (high $K_o$), which is called *high K stimulation* in electrophysiology.

## Cell membrane: a logarithmic operator for external signals

Not only the chemical senses such as smell and taste, but also almost all other sensations including vision and hearing are ultimately converted into electrochemical signals in the logarithmic fashion mentioned earlier. According to the above equation, the 3-dimensional concentration signals are converted into a 1-D membrane potential signal via a 2-D transducer, the cell membrane. Thereafter, the 1-D potential signal will trigger a sequence of physiological responses inside the cell.

## pH electrode: not an electrode but a battery

Acidity and alkalinity are also expressed logarithmically by pH, and the working principle of a pH electrode is just exactly the same as a membrane potential.

On the top of a pH electrode is a glass membrane permitting only the passage of protons, or more precisely, the hydronium ion ($H_3O^+$). The system is very similar to that used in measuring membrane potential except that the ionic current flows through

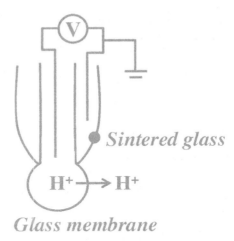

*Sintered glass*

$H^+ \longrightarrow H^+$

*Glass membrane*

a hole filled with sintered glass. The hole has no selectivity against ionic species but allows the free passage of ionic current without substantial potential drop. The potential measured with a voltmeter is therefore contributed mostly by the potential difference across the glass membrane. The whole system looks like a probe but is actually a concentration cell equipped with an ion selective membrane.

Similar to the previous approach, the membrane potential at 25°C can be calculated by:

$$V_m = \frac{RT}{nF} \ln \frac{H_o}{H_i} = \frac{59\,\text{mV}}{+1} [\log(H_o) - \log(H_i)] = -59\,\text{mV} \times \text{pH} + V^o$$

Where $H_o$ is the proton concentration to be measured in the test solution; $H_i$ is the inner proton concentration inside the glass membrane; $V_o$ is a constant potential which depends on $H_i$. The potential measured must be linearly correlated with the pH of test solutions.

## Impedance matching

One may be curious to know if it is possible to measure the potential output of a pH electrode with an ordinary voltage meter.

In a real test, the reading of the meter will always be "zero" regardless of the pH of the solution. The same situation will occur in trying to start up the engine of a car using dry cell batteries. Care must be taken to address the problem of impedance matching between the output and input signals.

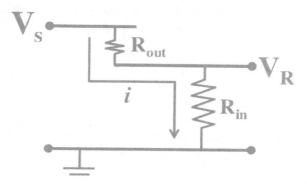

## Output *versus* Input Impedance

As revealed in the above figure, the input voltage signal ($V_s$) will certainly be "consumed" by its own output impedance ($R_{out}$). Only if the input resistance ($R_{in}$) is much higher than the output impedance ($R_{in} \gg R_{out}$), can the signal decay be neglected as in the following:

$$V_R = \frac{R_{in}}{R_{out} + R_{in}} V_s$$

$$V_R \cong V_s \quad \text{when} \quad R_{in} \gg R_{out}$$

The glass membrane of a pH probe is highly ion selective but possesses a high resistance of up to hundreds of $M\Omega$. To measure this kind of high resistance voltage signal, a voltmeter with higher input impedance is required. A pH meter is basically an FET-based voltmeter with high input impedance ($G\Omega$ order).

Electric eels inhabiting the Amazon can deliver a voltage output up to 600 V capable of stunning large animals like horses. On the other hand, the electric ray living in the ocean can only impose

30 V on the surroundings. Impedance matching gives at least part of the reason for this difference. The output impedance of the latter (sea water) is much lower than the former (fresh water). A voltage cannot be built up in such a low impedance environment without considerable leakage current. That may partially explain why the positive and negative poles of the electric organ of the ray are separated by their extended fin. The current leakage can be suppressed by the increased output impedance.

## Electronic conductors, ionic conductors and electrodes

On the surface of metals there are freely moving electrons to conduct electricity. Except for mercury, the common conductors are most often solid and called *electronic conductors* since the charge carriers are electrons. *Ionic conductors* are usually in liquid form such as sea water; the ions contained within serve as the charge carriers. *An electrode*, in the strict electrochemical definition, is the interface between an electronic conductor and an ionic conductor.

## The conductivity of ionic conductors

Since most ionic conductors are liquid, it is hard to define their electric resistances with a "solid" electric multi-meter. Since we use a measuring device based on solid electronic conductors, it is necessary to introduce an "electrode" to interface with the different charge carriers.

Let's start measuring the conductance (G) or resistance (R) of sea water by plunging two electrodes into the liquid. However, a steady reading cannot be obtained with a multi-meter.

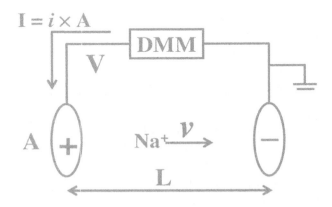

**Measuring Conductance with a Digital Multimeter**

The multi-meter imposes a voltage between the electrodes and measures the current; the resistance or conductance between the electrodes is determined by Ohm's law. Since the current between the two electrodes is ionic, the intensity will depend on the migration velocity ($v$) of ions along the electric field.

$$v = E \times u$$

where $u$ (the mobility of the ion) is the migration speed under unit electric field strength; $E$ is the electric field strength. Mobility is a constant dependent on the hydration radius, mass and charge of the ion and also the viscosity of the medium.

The flow of ions is better expressed as the flux ($j$), the number of particles passing through a unit area per unit time; and flux is just the product of the particle speed ($v$) and concentration ($C$). The current density ($i$) generated by the flux of an ion can be defined as the following:

$$j = Cv$$
$$i = nFj = nFCv = nFCEu$$

where $n$ and $F$ are the charge of the ion and Faraday's constant, respectively.

Assuming the electrodes have the same surface area ($A$) and the electric field is parallel, the electric field strength ($E$) will be linearly correlated with the voltage ($V$) by the distance ($L$) between electrodes. Moreover, the current intensity ($I$) is the product of the current density ($i$) and the area ($A$) of an electrode. The conductance ($G$) measured by this configuration will be:

$$V = EL$$
$$I = iA$$
$$G = \frac{1}{R} = \frac{I}{V} = \frac{iA}{EL} = \frac{i}{E} \times \left(\frac{A}{L}\right) = nFCu \times \frac{1}{\kappa} = G_0 \times \frac{1}{\kappa}$$

where $\kappa$ is the cell constant of the measuring "device," which is determined by the geometry of the "electrode set"; $G_0$ (conductivity) is the "intrinsic" conductance dependent on the properties of the ionic conductor including ion concentration ($C$), charge ($n$) and the mobility ($\mu$).

Since conductivity doesn't depend on the cell constant of the measuring device, the "normalized" conductance has become a very useful solution property giving the ease of conducting electricity. A commercial "conductivity electrode" is again not an electrode in the strict electrochemical definition; it is actually an electrode set with well-defined cell constant for converting conductance data to conductivity. The SI unit for conductance is the inverse of ohms ($\Omega$) and called Siemens (S); incorporated with the cell constant, the unit for conductivity is $Scm^{-1}$.

## Electroneutrality and shielding effects

Since conductivity ($G_o$) is directly proportional to the concentration (C), one may try to estimate ion concentration from the solution conductivity.

$$G_o = nFCu$$

However, for salt water (an aqueous solution of NaCl), the relationship holds only for concentrations below 10% (w/v). With higher concentrations, the conductivity will not increase as much as predicted.

You may have heard about the acidity and alkalinity of a solution, but have you heard that a solution could possess positive or negative charges? For a NaCl solution, we cannot just dissolve sodium ion into water, it would require an enormous amount of energy. To stabilize the solution, the cationic sodium ion (as the "center ion") must be surrounded by an "ionic atmosphere" (just like the electron cloud around a nucleus in quantum theory) of the anionic "counter ion" chloride, in this case, to achieve charge neutrality. Both macroscopically and microscopically, electroneutrality is therefore a universally observed phenomenon.

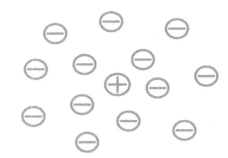

Center Ion and the Ionic Atmosphere
of its Counter Ions

The net charge of the center ion is now said to be "shielded" by the surrounding ionic atmosphere, and its migration under an electric field will be hampered. The higher the concentration (or more precisely, the ionic strength in Debye-Hückel theory), the stronger the shielding effect, which will lead to a decrease in ion mobility and conductivity.

$$\text{Ionic strength} = \frac{1}{2}\sum_i C_i Z_i^2$$

where $C_i$ and $Z_i$ are the concentration and charge number of the ion species $i$, respectively.

## Effect of thermal motion, dielectric constant on the microscopic world

Due to electroneutrality, a large body such as a human being can hardly carry a net charge as compared with its mass, i.e. the charge to mass ratio (Q/M) is almost zero. There are males and females, but no positive and negative "guys." Therefore, human beings are aware of gravity but not affected by electric fields.

Things become totally different in the microscopic world. From cells, bacteria, viruses, proteins to ions, the smaller the size, the stronger is the effect of an electric field. Particles with opposite charges are thought to aggregate by mutual electrostatic attraction, but the tendency is balanced by the thermal motion of charged particles at a temperature higher than absolute zero. In other words, thermal motion prevents "electrostatic collapse" or a "black hole" in the microscopic world.

The (electric) permittivity ($\varepsilon$) of the medium provides another "electrostatic protection" for the microscopic world:

$$F_e = \frac{q_1 q_2}{4\pi\varepsilon r^2}$$

where $F_e$ is the electrostatic (Coulombic) force between two particles with charges of $q_1$ and $q_2$; $r$ is the distance between the particles; $\varepsilon$, the permittivity of the medium, is the product of the permittivity in vacuum ($\varepsilon_o$) and the relative permittivity of the medium ($\varepsilon_r$).

$$\varepsilon = \varepsilon_0 \varepsilon_r$$

The relative permittivity of a medium ($\varepsilon_r$) is also called the dielectric constant of medium. Since there are no obstacles between two particles if they are put in a vacuum, the dielectric constant ($\varepsilon_r$) is 1 with a maximized electrostatic force. When they are immersed in a polar medium such as water, the electrostatic force will be damped by the induced dipoles of water molecules somewhat similar to the aforementioned shielding effect. The dielectric constant of water is about 70, i.e. the electrostatic force in such a medium will be reduced to about 1/70 of the value in vacuum. That is exactly why water is selected as "the medium of life"; the charged biomolecules such as DNA, protein will be stabilized in a medium with a high dielectric constant.

### Effect of hydration on ion mobility

In the investigation of conductivity, we found that the mobility of the potassium ion ($K^+$) is higher than that of the sodium ion ($Na^+$). One may be curious about why a cation with a larger radius and mass would be faster than a smaller one.

In aqueous solution, each sodium ion is hydrated with about 6 water molecules, but there are only 4 for the potassium ion. It is the hydrated molecular clusters that are moving in water, not the "naked ions."

### Measuring conductivity using an inverting amplifier

The signal output ($V_s$) from a conductometer to its probe is, surprisingly, an alternating voltage signal (2V, 2kHz) rather than a static one. The signal passing through the sample solution is to be processed via an inverting amplifier to obtain an AC output ($V_o$) with the amplitude directly proportional to the conductance ($G$) of the solution. The output signal can be rectified and normalized with cell constant ($\kappa$) to obtain the conductivity ($G_o$).

$$V_o = -V_s \times \frac{R_f}{R} = -V_s \times R_f \times G = -V_s \times R_f \times \frac{G_o}{\kappa}$$

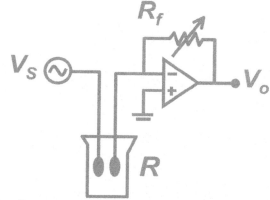

## Conductance Measuring Circuit

To increase the precision, a suitable resistor ($R_f$) should be chosen to obtain a reasonable output for measurement and calculation. Moreover, an OP-amp with high input impedance should be used to eliminate impedance matching problems when a solution with low conductivity is measured.

### Why use an AC signal to measure conductivity?

Surely we can measure conductivity using a DC voltage signal from a battery or power supply. If two stainless clips are used as the electrodes, you will find that bubbles (hydrogen gas) will evolve from the surface of the clip connected to the negative input. On the other side (the positive one), the stainless surface starts getting rusty. This is actually a reduction and an oxidation process occurring on the clip surfaces, which will disturb the measurement and change the cell constant. Using an AC voltage source can effectively suppress the redox process and electrode fouling to obtain a stable reading.

# Chapter 03

# Redox Potential

## Heterogeneous reactions on electrodes

In the previous experiment, we found out that the redox reactions (or formally, oxidation-reduction reactions) are confined to the surface regions of the clips. These reactions belong to one kind of heterogeneous reaction which occurs only at the interface between an electronic conductor and an ionic conductor, the electrode in the strict electrochemical definition. The electronic current (flow) will convert to an ionic current (flow) on the interface, and vice versa. The current related to this conversion process is called *Faradaic current*. In a more specific definition, *electrochemistry* is the chemistry concerned with heterogeneous processes occurring on an electrode.

## Ag/AgCl reference electrode

To facilitate a heterogeneous reaction, one can impose a voltage (i.e. energy) on the electrode, but the reproducibility is not satisfactory if we don't refer it to a standard potential. A reference electrode (e.g. the Ag/AgCl electrode) is again usually not an "electrode" in the strict electrochemical definition, but a device based on a reversible and quick electrochemical process that maintains thermodynamic equilibrium even with a considerable Faradaic current.

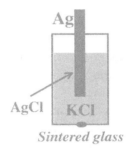

*Sintered glass*

## Ag/AgCl reference electrode

An Ag/AgCl electrode can be made by first slightly etching a silver wire with dilute nitric acid. While dipped into an aqueous KCl solution, the silver wire has a few positive volts imposed on it using a stainless or platinum bar as the counter electrode (ground). The silver wire will be electrochemically etched and a layer of black AgCl solid will deposit onto the silver surface:

$$Ag_{(s)} + Cl^- \rightarrow AgCl_{(s)} + e^-$$

Actually, the true story is:

$$Ag_{(s)} \rightarrow Ag^+ + e^-$$
$$Ag^+ + Cl^- \rightarrow AgCl_{(s)}$$

The silver is oxidized to become silver ion and then immediately interacts on the electrode surface with chloride ion to form AgCl precipitate. The "rusty" silver wire is then placed into a cell contained saturated aqueous KCl solution and sealed with sintered glass or agar to permit an ionic current without appreciable leakage. The concentration (or more accurately, the activity in Debye-Hückel theory) of chloride ion around the silver wire will not be altered much to maintain the following electrochemical balance:

$$Ag_{(s)} + Cl^- \leftrightarrow AgCl_{(s)} + e^-$$

In other words, the free electrons on the silver surface of an Ag/AgCl or other reference electrode will keep a stable electrical potential for referral during an electrochemical process.

## Redox potential, Nernst equation and formal potential

Membrane potential is generated as a natural response to cancel the trend for ion diffusion. Similarly, redox potential (or oxidation-reduction potential, ORP) is another natural response to counteract the trend of reduction or oxidation reactions.

$$O + ne^- \leftrightarrow R$$

where $O$ and $R$ are respectively the oxidized and reduced species in the reaction; $n$ is the number of charges transferred. Movement toward the R (right) side is called reduction; the change in free energy ($\Delta G$) will be:

$$\Delta G = \Delta G^o + RT(\ln[R] - \ln[O]) = \Delta G^o + RT \ln \frac{[R]}{[O]}$$

where $\Delta G^o$ is the difference in free energy at the standard state (25°C, 1 bar, 1M). In the derivation of membrane potential, $\Delta G^o$ is zero since the chemical species on both sides of the reaction is the same. In this case, $\Delta G^o$ cannot be neglected.

The following experiment is conducted using platinum as the working (indicating) electrode, and the open circuit voltage is measured by referring to an Ag/AgCl reference electrode.

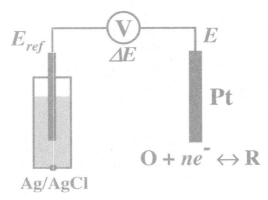

**Open Circuit Voltage Between Electrode**

The electrons for the reduction (with potential $E$) are supplied from the Ag/AgCl electrode (with potential $E_{ref}$), so the change in electric energy will be:

$$-nF(E - E_{ref}) = \Delta G^o + RT \ln \frac{[R]}{[O]}$$

The energy change will be balanced by the free energy change mentioned earlier (the right terms). The "electro-chemical" balance can be rearranged to obtain the Nernst equation:

$$E - E_{ref} = \Delta E = E^{o'} + \frac{RT}{nF} \ln \frac{[O]}{[R]}$$

where $E^{o'}$ is the "formal potential" of the redox system.

### Reduced species, reducing species, and reductants

It may be somewhat confusing to talk about reduced and reducing species in a redox reaction. The oxidized species is a chemical which has been oxidized and goes to a higher oxidation state. Chemicals in higher oxidation state may accept electrons to return to their original reduced format. In the same way, a reductant will be oxidized to produce electrons to complete "Kirchhoff's cycle." As in the following half and whole reactions:

$$Fe(III) + e^- \leftrightarrow Fe(II)$$

$$\text{Reductant} \leftrightarrow Oxidant + ne^-$$

$$Fe(III) + \text{Reductant} \leftrightarrow Fe(II) + Oxidant$$

The oxidized species, ferricyanide: *Fe(III)*, will oxidize some reductants if the above reaction does occur. So, an oxidized species may have an oxidizing power and may become an oxidizing agent (or species) or the oxidant in a redox reaction. Similarly, a reduced chemical may become a reducing agent and thus the reductant in a redox system.

## pH buffers and ORP buffers

pH buffers such as a phosphate buffer are frequently used in bio-experiments. The concentration of proton (and therefore the pH of the solution) can be controlled within a range around the dissociation constant (the pKa) of the following reversible (de)protonation reaction:

$$H_2PO_4^- \xleftrightarrow{Ka} H^+ + HPO_4^{2-}$$

$$[HA \xleftrightarrow{Ka} H^+ + A]$$

The pH can be calculated by the balance in free energy as follows:

$$G_{HA}^o + RT\ln[HA] = G_H^o + RT\ln[H] + G_A^o + RT\ln[A]$$

$$\ln\frac{1}{[H]} = \frac{G_H^o + G_A^o - G_{HA}^o}{RT} + \ln\frac{[A]}{[HA]}$$

$$pH = pKa + \log\frac{[HPO_4^{2-}]}{[H_2PO_4^-]}$$

To prepare a phosphate buffer with a desired pH, we can just tune the molar ratio of $KH_2PO_4$ and $Na_2HPO_4$ salt dissolved in the solution. Similarly, based on the Nernst equation, we can prepare an "ORP buffer" with a desired redox potential simply by dissolving a suitable amount of oxidized (ferricyanide, $K_3Fe(CN)_6$, as in the example) and reduced species (ferrocyanide, $K_4Fe(CN)_6$) in a solution (e.g. 0.1M phosphate buffer).

$$Fe(III) + e^- \leftrightarrow Fe(II)$$

$$E = E^{o\prime} + \frac{RT}{nF}\ln\frac{[Fe(III)]}{[Fe(II)]}$$

## Scrambling for electrons/protons on an ORP/pH electrode

ORP and pH electrodes are actually devices used for measuring the tendency of physical phenomena, redox and diffusion processes, respectively, to occur. The diffusion of protons across the proton selective membrane of a pH electrode is counteracted by the dragging force of electrical migration, just like the scrambling for a proton on the membrane.

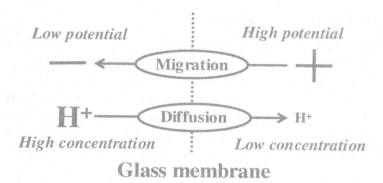

**Glass membrane**

**Balancing Diffusion and Migration "Force" on the Glass Membrane of a pH Electrode**

In the case of the ORP electrode, it is the free electrons on the electrode surface that are scrambled. When the concentration of oxidized species, say ferricyanide, is higher than its reduced form (ferrocyanide), the ferricyanide will tend to pick up an electron from the electrode surface and be transformed into its reduced form, ferrocyanide. In this case, the process will be counteracted by the elevated potential of the electrode which lowers the "escaping" energy of the electrons on the surface. As the result, the potential measured will be more positive than the formal potential to prevent the "escape" of the electrons.

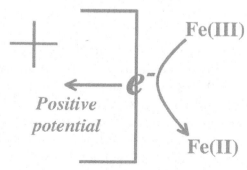

**Electrode with Positive Potential Traps Electrons to Resist Reduction.**

## An ORP electrode is still a battery

Both ORP and pH electrodes read the "open circuit voltage" between the two voltage outputs of the electrode. Of course the voltage had better be measured with a voltmeter with high input impedance. If a small resistor is connected between the outputs, a tiny current will appear. Now, the pH electrode is working as a concentration cell, or concentration "battery", and the ORP electrode works just the same as an ordinary chemical battery.

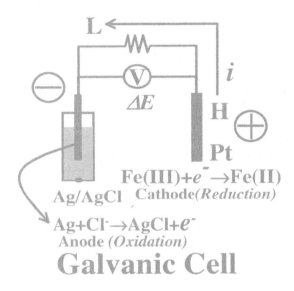

Galvanic Cell

In the above scheme, with the concentration of ferricyanide sufficiently higher than ferrocyanide, electrons on the platinum working electrode will be drawn by the reduction process, and the platinum electrode conducting a reduction reaction is called the "cathode" of the electrochemical system. The *cathode* was named for its tendency to accumulate cations, but now it is defined as the electrode on which the reduction reaction occurs.

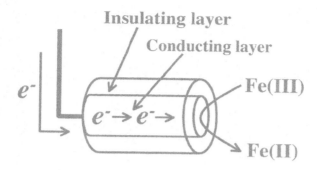

# Reduction on an Electrode Surface

Based on Kirchhoff's rule, the reduction process will lead to an electron flow from the Ag/AgCl electrode to the platinum electrode. The "electron supply" for the left side, Ag/AgCl, now turns out to be the oxidation reaction on its surface, and the Ag/AgCl electrode conducting an oxidation reaction is called the "anode" (the place gathering anions). Consequently, the electronic current flows from the platinum cathode (+: high potential) to the Ag/AgCl electrode (-: low potential). Beneath the solution is the ionic current which flow froms low (Ag/AgCl) to high potential (platinum). With this electrochemical configuration, the chemical energy of the redox system is converted into electricity, and the system is called *a Galvanic cell.* The cathode of a Galvanitc cell (battery) will be the positive output, exactly as we've learned in our high school chemistry. So, the ORP electrode is actually a chemical battery.

## Electrolytic cell: driving chemical reactions with electricity

In the following, we now impose a voltage on the system from a DC power supply:

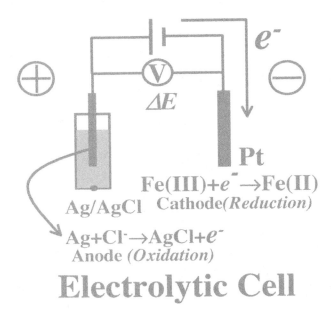

$$Fe(III)+e^-\rightarrow Fe(II)$$

Ag/AgCl  Cathode(*Reduction*)

$$Ag+Cl^-\rightarrow AgCl+e^-$$

Anode (*Oxidation*)

# Electrolytic Cell

The high energy electrons gushing out from a DC power supply drive the reduction reaction on a platinum electrode even if the concentration of ferricyanide is low. Therefore, the electronic current is still going from the platinum cathode to the Ag/AgCl anode. However, the cathode of this system (an electrolytic cell) is the negative pole, the opposite of a Galvanic cell. In an electrolytic cell, electric energy is converted into chemical energy.

# Amperometric Measurement

## The Faraday constant is a large constant

Electrolytic and electroplating industries are highly energy consuming; for example, to obtain 9 g (one third of a mole) of aluminum requires 96500 coulombs of electricity (the Faraday constant).

$$Al^{3+}{}_{(aq)} + 3e^- \rightarrow Al_{(S)}$$

From the viewpoint of a human being, the Faraday constant is quite large, which will result in a large amount of energy being needed to convert chemicals by electrochemical means. However, this is certainly good news to a chemical analyst since minute amounts of chemicals will result in a considerable charge transfer or redox current. Amperometric (current) measurement is therefore a very sensitive method for chemical analysis.

### Two-electrode amperometric configuration

For the conveniences of signal conditioning, current signals are usually converted into voltage signals as follows:

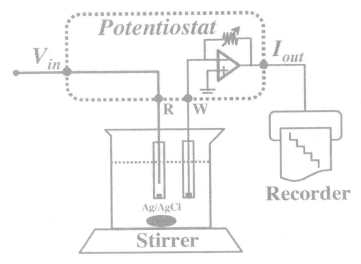

**Amperometric Method: Two-Electrode Configuration**
**R: reference electrode; W: working electrode**

The working electrode (W) is the one where the objective redox reaction occurs. W is connected to the negative input terminal of an OPA, and the negative feedback loop of the inverting amplifier configuration maintains a virtual ground potential for W. The voltage used to trigger a reaction ($V_{in}$) is actually imposed on the reference electrode (R), Ag/AgCl in this case. Consequently, if you impose a negative voltage on R, W will be positive to R, and an oxidation reaction may occur on W as described later in this chapter.

The working electrode and reference electrode are now respectively conducting the oxidation and reduction reactions to compensate according to Kirchhoff's rule. The current will flow from the negative (or inverting) input terminal to W (electronic current), then W to R (ionic current) and finally R to the voltage source (electronic current).

The current will bypass the high input impedance of the OPA and choose an easier way, the negative feedback loop with lower resistance ($R_f$), and the output voltage ($V_{out}$) will be proportional to the current ($I_{out}$):

$$V_{out} = R_f \times I_{out}$$

A current-measuring electrochemical instrument with a tunable potential-control is called a "potentiostat."

## The drawbacks of a two-electrode configuration

However, the two-electrode configuration cannot tolerate a large current, especially a Faradaic current, for the following reasons:

- A long-term voltage bias may deprive the reference electrode of its chemicals, such as the AgCl layer on the silver wire ($AgCl + e^- \rightarrow Ag_{(S)} + Cl^-$). This will lead to an error in reference potential.
- Due to Kirchhoff's rule, the transient current will be restricted by the current capacity of the reference electrode if it is not sufficient to compensate for the redox reaction on the working electrode.

The above troubles can be always overcome by using a reference electrode with higher capacity, but we have a cleverer solution.

## Three-electrode amperometric configuration

The major difference between two and three-electrode configurations is in the reference side.

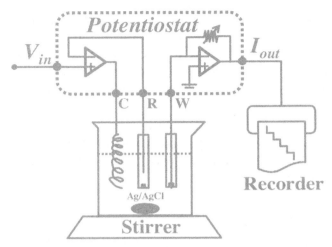

## Amperometric Method: Three-Electrode Configuration
R: reference electrode; W: working electrode; C: counter electrode

The reference electrode (R) is connected to the negative input of another OPA; the OPA protects the electrode with its inherent high input impedance and provides a voltage comparison function for the required voltage control. A relatively "weak" negative feedback exists via the connection of R (reference electrode) with C (counter electrode) and the electrolytes in the solution. Due to the virtual short between the positive and negative input of the OPA, the potential of R will practically be equal to $V_{in}$. The potential of the virtually grounded working electrode (W) is therefore "$-V_{in}$" with respect to the reference point, $-V_{in}$ versus Ag/AgCl in this case.

Although the potential is not exactly the same as R, the counter electrode (C), sometimes called the auxiliary electrode, now "comes to the rescue" for R. Instead of R, as in the two-electrode configuration, it is C that will perform the "mirror" redox reaction for W. The upgraded configuration can stand for prolonged use and a higher current.

## Supporting electrolytes

In an electrochemical experiment, supporting electrolyte or pH buffer at a concentration of 10–50 mM or so is normally added into the solution, which will have the following merits:

- The potential drop of the ionic conductor (IR drop) will be reduced as described later. The electrolytic process will be more focused on chemical conversion with less heating of the solution, i.e. higher energy efficiency.
- The shielding effect of the electrolyte can reduce the unwanted migration of some important ions caused by the voltage between working and counter or reference electrode.
- The shielding effect of the electrolyte can protect the electrode surface from the electrostatic adsorption of some interfering substances such as proteins.

## Potential drop across the ionic conductors

After adding the supporting electrolytes, one may think that the resistance of the ionic conductor (the solution) will be reduced or eventually shorted. How can we impose a voltage on a shorted circuit? The problem should be reconsidered more systematically.

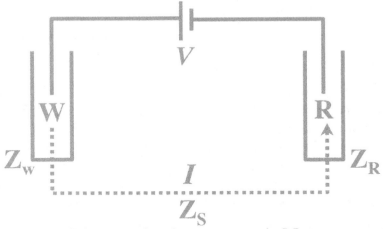

**Impedance Diagram for Amperometric Measurement**
$Z_W$: Interfacial impedance on working electrode
$Z_S$: Solution resistance
$Z_R$: Interfacial impedance on reference electrode

Not only does the solution have a resistance $(Z_S)$, but also the interfaces between the electrodes and the solution $(Z_W$ and $Z_R)$. Surely the interfacial ones will have some capacitive effect in addition to the resistive effect, so it is more accurate to use the term impedance. The potential drop will be a function of these impedances.

$$V = I \times (Z_W + Z_S + Z_R)$$

The reference electrode is generally designed with a fast and reversible redox reaction, so its impedance is much smaller than that of the working electrode $(Z_R \ll Z_W)$. In the case of the Ag/AgCl electrode, the surface of the silver wire was roughed by etching with thin nitric acid. The enlarged metal surface along with the porous precipitates of AgCl particles and the saturated KCl solution further enhance the reaction rate and therefore reduce the impedance. The impedance and thus the potential drop on the "reference side" will be much smaller than the "working side," as illustrated below.

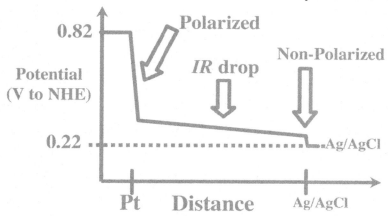

## Potential Profile of Amperometric Experiment
## Pt : platinum electrode

In the figure, +0.6V is imposed on a platinum electrode with respect to an Ag/AgCl electrode. NHE represents the normal hydrogen electrode. The platinum working electrode is said to be positively polarized due to the clear potential drop. The impedance of the Ag/AgCl reference electrode is low and not easily polarized with a significant potential drop. Therefore, the reference electrode is generally "non-polarizable." Ideally, the following relation holds for an amperometric experiment:

$$V \approx I \times Z_W$$

For an experiment in an organic solvent where electrolytes can hardly be dissolved, the distance between the working and reference electrodes should be made as small as possible to reduce the IR drop.

### The reason for using an Ag/AgCl electrode in a pH probe

We can find that both sides of the glass membrane of a pH probe are often using Ag/AgCl electrodes. Due to the low impedance of an Ag/ AgCl electrode, the potential profile will be:

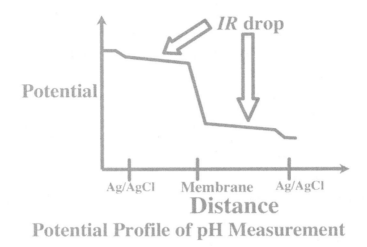

**Potential Profile of pH Measurement**

The majority of the voltage output of a pH probe will be contributed by the potential drop across the ion selective membrane, which is exactly the desired outcome of a successful experimental setup. The potential profile of an ORP probe will be very similar to the one for the previous amperometric measuring system.

**A counter electrode should be as large as possible**

With a 3-electrode configuration, the potential can be accurately imposed on the working electrode, but we should concern more about the counter electrode if the current and energy efficiency are to be considered.

$$V = I \times (Z_W + Z_S + Z_C)$$

If the impedance of the counter electrode ($Z_C$) and the solution resistance are small enough, almost all of the electric work is going to polarize the working electrode and conduct the electrochemical conversion.

$$P \approx I^2 \times Z_W$$

The simplest way to reduce $Z_C$ is to increase the surface area of the counter electrode.

## Amperometric vs. potentiometric signal

The current signal of an amperometric measurement is directly related to the concentration of analyte rather than its logarithmic value as obtained potentiometrically, say by pH and ORP. This kind of "linear" signal is narrow in dynamic range but can discriminate a tiny concentration change of the analyte.

The "logarithmic wide-range" potentiometric signals are suitable for "sensing" fluctuating environmental factors such as pH, taste and smell-related chemicals. The "sensitive" amperometric signals are of merit in "measuring" some physiological status indicators such as temperature, blood sugar for a delicate control.

## Why a positive electrode facilitates oxidation

In an experiment imposing +0.6V (vs. Ag/AgCl) on a platinum electrode, hydrogen peroxide added into the solution will be oxidized on the surface of the positively polarized working electrode.

$$H_2O_2 \rightarrow O_2 + 2H^+ + 2e^-$$

To trigger the reaction, we can increase the concentration of the reactant (hydrogen peroxide) or decrease the product (oxygen molecule). Or we can increase the solution pH to suppress the proton concentration. However, these will be either inefficient or not suitable for bio-concerns since pH control is very important.

The best remedy is to decrease the "escaping" energy of the electron on the electrode surface, which can be easily done by imposing a positive potential. We will deal with this problem in a more quantitative manner in the next chapter.

# Chapter 05

# Steady-State Voltammetry

## Tuning the energy barrier using electric potential

$$O + ne^- \rightarrow R$$

In the above reaction, we cannot ascertain whether the oxidized species, O, is capturing $n$ electrons or if R is sending $n$ electrons to O. The symmetry factor ($\alpha$) is introduced as a compromise in this trade-off.

$$O + \alpha_c ne^- \leftrightarrow R - \alpha_a ne^-$$
$$\alpha_c + \alpha_a = 1$$

$\alpha_c$ is the cathodic (reductive) symmetry factor, and $\alpha_a$ is the anodic (oxidative) symmetry factor. The energy diagram will become:

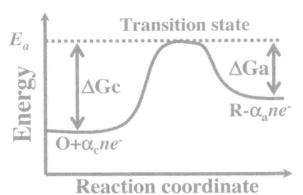

**Energy Diagram of Electrochemical Reaction**
$E_a$: activation energy

In reality, the energy barrier depends not only on the free energy difference ($\Delta G_c$ or $\Delta G_a$), but also on the electron energy. For the reductive forward reaction, the higher (the more positive) the electric potential, the lower the energy of the electrons and the higher the energy barrier.

$$E_a(R) = \Delta G_c + \alpha_c nFE$$

On the other hand, the energy barrier for the oxidative backward reaction will be:

$$E_a(O) = \Delta G_a - \alpha_a nFE$$

As a distinct character of electrochemical reactions, the energy barrier can be tuned easily by adjusting the potential.

**Potential-controlled reaction kinetics**

The rate constant (k) for a reaction is generally related to the activation energy ($E_a$) and absolute temperature (T) as an Arrhenius expression.

$$k = A e^{\frac{-E_a}{RT}}$$

where $A$ is the Arrhenius constant for the reaction.

After incorporating the corresponding Arrhenius constant ($A_c$ and $A_a$), the rate constant for a reductive ($K_c$) and oxidative reaction ($K_a$) will be:

$$K_c = A_c e^{\frac{-(\Delta G_c + \alpha_c nFE)}{RT}} = K_c^o e^{-\alpha_c nFE / RT}$$

$$K_a = A_a e^{\frac{-(\Delta G_a - \alpha_a nFE)}{RT}} = K_a^o e^{\alpha_a nFE / RT}$$

According to the Nernst equation, describing equilibrium status:

$$E = E^{o'} + \frac{RT}{nF} \ln \frac{[O]}{[R]}$$

At formal potential ($E^{o'}$), the reduction and oxidation reaction (rates) will balance each other when the concentrations of O and R are equivalent.

$$E = E^{o'}$$
$$Rate = K_c[O] = K_a[R]$$
$$\because [O] = [R] \therefore K_c = K_a = K_o$$

Therefore, the expressions for the rate constants can be simplified using a united rate constant $K_o$ which does not depend on the potential:

$$K_c = K_o e^{-\alpha_c nF(E - E^{o'})/RT}$$

$$K_a = K_o e^{\alpha_a nF(E - E^{o'})/RT}$$

The net oxidative (anodic) reaction rate ($R_a$) can be expressed as:

$$R_a = K_a[R] - K_c[O]$$

$$= K_o\{[R]e^{\alpha_a nF(E - E^{o'})/RT} - [O]e^{-\alpha_c nF(E - E^{o'})/RT}\}$$

A redox reaction is therefore a dynamically coupled oxidative and reductive reaction with kinetics controllable by the electrical potential.

## Rate of heterogeneous reaction and flux

Care must be taken for the heterogeneous nature of an electrochemical reaction; the units for the reaction rate are different from the homogeneous ones. The rate of a first order homogeneous reductive reaction ( $R_{homo}$ ) can be expressed as:

$$R_{homo} = -\frac{d[O]}{dt} = k_{homo}[O]$$

The unit of the rate constant ( $k_{homo}$ ) is the reciprocal of time ($S^{-1}$). The reaction rate ( $R_{hetero}$ ) of a first order heterogeneous reductive reaction should be expressed using the concept of flux.

$$R_{hetero} = -\frac{1}{A}\frac{dN_o}{dt} = k_{hetero}[O] = Flux$$

Within a unit surface area ($A$) on the electrode, there will be a number of oxidized species appearing ($dN_O > 0$) or disappearing ($dN_O < 0$) per unit time ($dt$). Therefore, flux has SI units of $mol \cdot m^{-2} \cdot S^{-1}$. The SI unit for the rate constant will be $mS^{-1}$.

## Flux and current density

Since it is the flux ($j$) of redox species, not the total flow that is related to the heterogeneous kinetic parameters, the total flow we measured, the current ($I$), had better be converted into its flux-related quantity, the current density ($i$) for subsequent discussion.

$$nFj = i = \frac{I}{A}$$

## Butler-Volmer equation and the surface concentration

The above concepts are included in the following Butler-Volmer equation:

$$I = nFAj = nFAK_o\{[R]e^{\alpha_a nF(E-E^{o'})/RT} - [O]e^{-\alpha_c nF(E-E^{o'})/RT}\}$$

For a fast reaction ($K_o \to \infty$) with limited current (I), the final term should be zero:

$$[R]e^{\alpha_a nF(E-E^{o'})/RT} - [O]e^{-\alpha_c nF(E-E^{o'})/RT} = 0$$

However, it is common to add either O or R into the solution, so this term should not be cancelled to zero as expected. There must be some contradictions.

In fact, rather than using the bulk (average) concentrations for [R] and [O], it is the concentrations adjacent to the electrode surfaces that will determine the kinetics. The previous equations should be modified to be:

$$I = nFAj = nFAK_o\{[R]* e^{\alpha_a nF(E-E^{o'})/RT}$$
$$-[O]* e^{\alpha_c nF(E-E^{o'})/RT}\}$$
$$[R]* e^{\alpha_a nF(E-E^{o'})/RT} - [O]* e^{-\alpha_c nF(E-E^{o'})/RT}$$
$$= 0 \text{ (for a reversible reaction)}$$

where $[R]^*$ and $[O]^*$ represent the surface concentrations of R and O, respectively. The second equation can be rearranged into the following alternative format of the Nernst equation which will hold even if considerable current is flowing:

$$E = E^{o'} + \frac{RT}{nF}\ln\frac{[O]*}{[R]*}$$

43

## Reverse operation of an ORP system with a concentration gradient

During the measurement of redox potential, the potential is measured with different molar ratios of [O]/[R]. In an amperometric measurement with constant potential (steady state voltammetry), the molar ratio on the electrode surface, [O]*/[R]*, is controlled by the imposed potential. With a considerable redox current, a steady concentration gradient will be generated near the electrode surfaces.

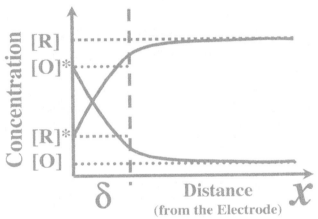

## Concentration Profile during Oxidation

The situation is said to be in "steady state" but not in "equilibrium." Of course, if the current density is low enough, the concentration gradient will be negligible as follows:

$$E = E^{o'} + \frac{RT}{nF} \ln \frac{[O]*}{[R]*} = E^{o'} + \frac{RT}{nF} \ln \frac{[O]}{[R]}$$

# A reference electrode: an electrochemical rectifier

A reference electrode is usually built around a fast reversible reaction with a large rate constant. A small overpotential (a potential deviating from equilibrium) making:

$$[R]e^{\alpha_a nF(E-E^{o'})/RT} - [O]e^{-\alpha_c nF(E-E^{o'})/RT} \neq 0$$

will result in considerable redox current. The electrode will act as an electrochemical rectifier with the following characterization curve:

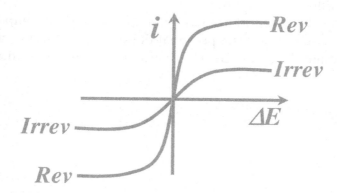

**Characterization Curve of Reversible (*Rev*) and Irreversible (*Irrev*)Electrode Process**

Mounted in a stable environment (such as Ag/AgCl in saturated KCl solution), the potential of a reference electrode tends to maintain a certain thermodynamically defined level even if considerable current is flowing.

## Fick's laws and the diffusion layer

A steady flux ($j$) with a steady concentration gradient can be described by Fick's first law; and the flux through the electrode is thus related to the concentration profile adjacent to the electrode surface.

$$j = -D\frac{dC}{dx}\Big|_{x=0} = -D\frac{[R]-[R]^*}{\delta}$$

where $D$ is the diffusion coefficient for the substance in the medium at a certain temperature (viscosity); $x$ is the distance from the electrode surface; $\delta$ is the thickness of the hypothetical diffusion layer on the electrode within which the concentration profile is built up. Information about $\delta$ is very helpful in calculating flux and consequently the current density.

In a steady state, $\delta$ is a constant, but it will become a complicated function of time when the potential doesn't hold constant during an amperometric measurement (transient voltammetry in the next chapter).

# Chapter 06

# Transient Voltammetry

## Potential step: Cottrell experiment

As an extreme example, the potential starts from a point at which no reaction occurs and instantly changes (steps) to a positive potential at which an oxidative reaction occurs at full speed. At this moment, R will soon be depleted on the electrode ($[R]^* = 0$ when $t > 0$).

$$R^* \rightarrow O^* + e^- \text{(Potential } \uparrow, \text{Energy of electron } \downarrow\text{)}$$

The concentration profiles of R near the electrode surface will become:

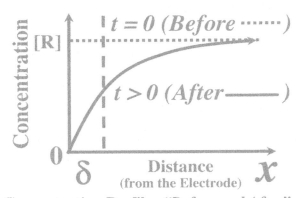

**Concentration Profiles** *"Before and After"*
**a Positive Potential Step of Sufficient Voltage**

Since the rate of oxidation is sufficiently quick, the oxidation current will be limited (controlled) merely by the diffusion process (mass transfer) to the electrode surface. The remaining task is to solve the time-dependent diffusion equation (Fick's second law) with three boundary conditions.

$$\frac{\partial C}{\partial t} = D\frac{\partial^2 C}{\partial x^2}$$

### Boundary conditions

1. $t = 0$, $C_{0 \sim \infty} = [R]$: Before the potential step, the concentration at anywhere from the electrode surface ($C_0$: $t = 0$, $x = 0$) to a distance sufficiently far away from the electrode ($C_\infty$: $t = 0$, $x = \infty$) will be equal to the bulk concentration, $[R]$.

2. $t \geq 0$, $C_\infty = [R]$: At a distance sufficiently far away from the electrode surface, the concentration will always be the bulk concentration.

3. $t > 0$, $C_0 = [R]^* = 0$: Immediately after the potential step, the concentration of R at the electrode will soon become zero.

Since the concentration is both a function of time and distance, the dynamic diffusion equation is certainly not as simple as it looks.

The concentration first has to be normalized and substituted into the second law.

$$\gamma = \frac{C - C_\infty}{C_\infty}$$

$$\frac{\partial \gamma}{\partial t} = D \frac{\partial^2 \gamma}{\partial x^2}$$

The equation is transformed with respect to time using a Laplace transformation:

$$\frac{\partial \gamma}{\partial t} \xrightarrow{\ L(t)\ } S_t \overline{\gamma} - \gamma(0)$$

$$D \frac{\partial^2 \gamma}{\partial x^2} \xrightarrow{\ L(t)\ } D \frac{\partial^2 \overline{\gamma}}{\partial x^2}$$

The merit of normalization is now obvious since the concentration will be equal everywhere at $t = 0$ (the first boundary condition), and $\gamma$ is therefore zero. Consequently, the balance in the $S_t$ domain is simplified to:

$$S_t \overline{\gamma} = D \frac{\partial^2 \overline{\gamma}}{\partial x^2}$$

The subscript $t$ in the "$S_t$" domain indicates the transformation was conducted with respect to time. The next step is to transform the equation in the $S_t$ domain with respect to $x$ and then invert the $S_x$ domain equation to solve the above equation. After introducing the second and third boundary conditions, the $S_t$ domain transform can be solved to remove the $x$ derivatives.

$$\overline{\gamma} = \frac{-1}{S_t} \exp[-(S_t/D)^{\frac{1}{2}} x]$$

Finally, the concentration equation can be obtained by inverting the above transform with respect to time.

$$C = C_\infty \, erf\left(\frac{x}{2\sqrt{Dt}}\right)$$

Concentration is therefore the error function of $x$ and $t$. An error function is defined as:

$$erf(z) = \frac{2}{\sqrt{\pi}} \int_0^z \exp(-t^2)dt$$

The numerical range of the output of an error function will be $0 \le erf(z) \le 1$, so the concentration can be normalized to obtain the following profile:

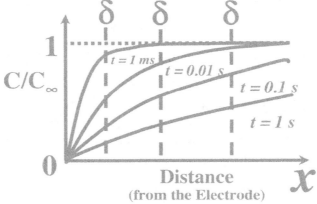

**Normalized Concentration Profiles During a Positive Potential Step of Sufficient Voltage**

As revealed by the profiles, the thickness of the diffusion layer will increase with time and the redox current will certainly decrease. However, it is hard to obtain the concentration gradient at the electrode directly from the dynamic profile, so the derivation goes back to the "resolved $S_t$ domain transform" and then the differential against $x$ is taken:

$$\bar{\gamma} = \frac{-1}{S_t}\exp[-(S_t/D)^{\frac{1}{2}}x]$$

$$\frac{\partial\bar{\gamma}}{\partial x} = (S_t D)^{-\frac{1}{2}}\exp[-(S_t/D)^{\frac{1}{2}}x]$$

It is the concentration gradient on the electrode surface ($x = 0$) that is important for us, so we put $x = 0$ and then invert the transform with respect to time.

$$\frac{\partial \overline{\gamma}}{\partial x} \xrightarrow{L^{-1}(t)} \frac{\partial \gamma}{\partial x}$$

$$(S_t D)^{-\frac{1}{2}} \exp[-(S_t/D)^{\frac{1}{2}} x]|_{x=0} = (S_t D)^{-\frac{1}{2}} \xrightarrow{L^{-1}(t)} \frac{1}{\sqrt{\pi Dt}}$$

$$\left(\frac{\partial \gamma}{\partial x}\right)_{x=0} = \frac{1}{\sqrt{\pi Dt}}$$

The time-dependent redox current can be obtained by substituting into Fick's first law:

$$I(t) = nFAj(t) = nFAD\left(\frac{\partial C(t)}{\partial x}\right)_{x=0} = nFAD\left(\frac{\partial \gamma(t)}{\partial x} C_\infty\right)_{x=0}$$

$$I(t) = nFAD\frac{C_\infty}{\sqrt{\pi Dt}} = nFAD\frac{C_\infty - 0}{\delta}$$

$$\delta = \sqrt{\pi Dt}$$

As predicted by the error function, the thickness of the diffusion layer ($\delta$) will increase with time, but the above Cottrell equation gives a more quantitative result. As a consequence, the resulting current after the potential step is decaying with the inverse of the square root of $t$.

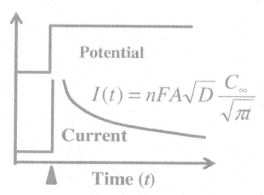

**Current Response of a Potential Step**

This kind of dynamic data is very useful in obtaining the information about the number of charges transferred $n$, the theoretical surface area of the electrode $A$ (usually larger than the geometrical one) and also the diffusion constant $D$.

**Potential ramp**

Similar to the starting conditions of the Cottrell experiment, the potential is now linearly swept at a constant speed $(v)$; the current signal $(I)$ can be discriminated as a capacitive current to charge the electrode surface $(I_c)$ and a Faradaic current that converts ionic and electronic flow $(I_F)$.

$$I = I_C + I_F = C_P \frac{dE}{dt} + I_F = C_P v + I_F$$

Since the capacitance of the electrode $(C_p)$ is constant, a linear sweep experiment has the merits of a constant capacitive current $(C_p v)$ to ease the data subtraction. However, the solution of its dynamic diffusion equation (Fick's second law) is even more complicated than for the Cottrell experiment. Actually, we cannot obtain an analytical result, and a numerical method or computer simulation is required. With some imagination, let's "simulate" a potential ramp experiment as a "slower" potential step experiment as follows:

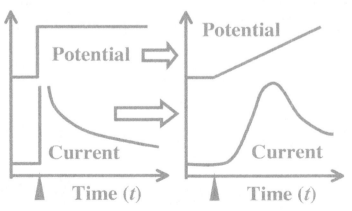

**From Potential Step to Potential Ramp**

With increasing scan rate ($v$), the peak height ($I_p$) will increase without changing the potential of the maximum current ($E_p$).

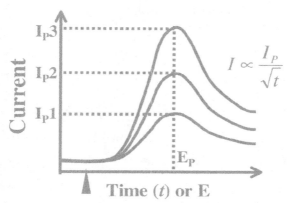

**Potential Ramps with Different Scan Speed**

Or more precisely, $I_p$ is proportional to the square root of $v$, and $E_p$ is independent of $v$.

$$E_P = \left( E^{O'} + \frac{RT}{nF} \ln\sqrt{D_O/D_R} \right) + \left( \frac{RT}{F} \right)\frac{1}{2n} \times \frac{1}{\log e}$$

$$= E^{\frac{1}{2}} + \left( \frac{RT}{F} \right)\frac{1}{2n} \times \frac{1}{\log e}$$

$$E_P \cong E^{O'} + \left( \frac{RT}{F} \right)\frac{1}{2n} \times \frac{1}{\log e}(\text{when} D_O \cong D_R)$$

$$I_P \propto n^{\frac{3}{2}} A C_\infty \sqrt{Dv}$$

where $E^{1/2}$ is called the half wave potential at which the current value is about half of $I_p$. For the current at a potential higher than $E_p$, similar current decay as in the Cottrell experiment appears with the increasing thickness of the diffusion layer.

## Cyclic voltammetry: the most popular experiment

After some "Cottrell decay", the potential is then swept back to the starting point to complete a cyclic scan. The following shows the potential change during three complete cycles:

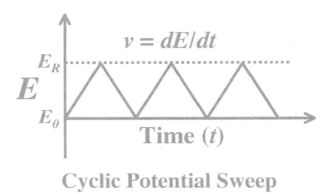

**Cyclic Potential Sweep**

where $E_O$ is the starting potential of the forward sweep; $E_R$ is the reversal potential of a sweep backwards. The resulting IV curve (voltammogram) will be:

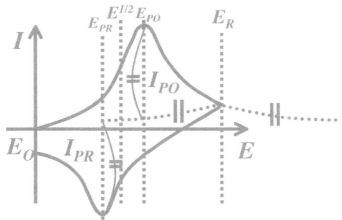

**Cyclic Voltammogram of a Reversible System**

where $E_{PO}$ and $E_{PR}$ are the potentials of oxidative and reductive peaks, respectively; $I_{PO}$ and $I_{PR}$ are the corresponding oxidative and reductive peak currents. These important parameters of the cyclic voltammogram (or CVgram) have the following relationships:

$$E_{PO} = E^{\frac{1}{2}} + \left(\frac{RT}{F}\right)\frac{1}{2n} \times \frac{1}{\log e}$$

$$E_{PO} - E^{\frac{1}{2}} \cong E^{\frac{1}{2}} - E_{PR} \cong \frac{RT}{2nF} \times \frac{1}{\log e}$$

$$|I_{PO}| \cong |I_{PR}|$$

Again, with some imagination, the concentration profile during a cyclic potential scan will be as follows:

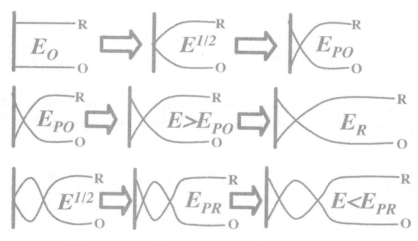

## Concentration Profiles in CV of a Reversible Electrochemical Process

The molar ratio of R/O at the electrode surface, i.e. [R]*/[O]*, is exactly controlled by the imposed potential, for example, [R]*/[O]* = 1 at $E^{1/2}$.

### Between reversible and irreversible process

Once the scan speed has been increased to certain level, the redox reaction rate and also the diffusion are unable to keep up with the speed of potential change. The $I_p$ will appear at higher $E_p$ and be smaller than predicted by the scan rate.

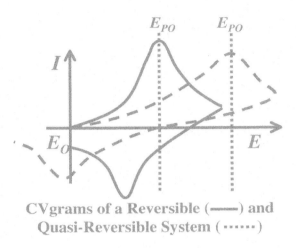

**CVgrams of a Reversible (——) and Quasi-Reversible System ( ······ )**

Therefore, the reversibility of a redox reaction for a CV experiment will depend on the scan rate as in the following plot:

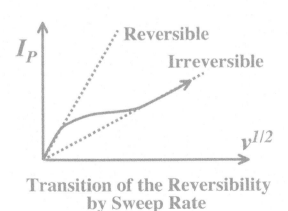

**Transition of the Reversibility by Sweep Rate**

Between the reversible and irreversible process is the quasi-reversible one. For a "pure" irreversible process, O will not be produced to form a reductive peak as the following:

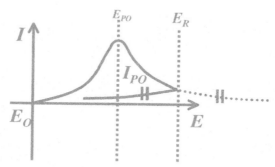

## CVgram of an Irreversible System

The current in the backward scan is just the "mirror" signal of "Cottrell decay".

Not only valuable for obtaining some useful parameters about an electrochemical process, cyclic voltammetry (CV) is also very useful for detecting unknown redox species which may interfere with an investigation. Therefore, CV is among the most popular electrochemical methods for biomedical research.

## Equivalent circuit for electrode processes

Information and phenomena occurring on an interface can be efficiently investigated by impedance analysis giving the resulting equivalent circuit. The following is a typical model circuit for an electrode process in which the elements for a Faradaic process and a non-Faradaic process (the charging and discharging processes of electrode capacitance) are connected in parallel.

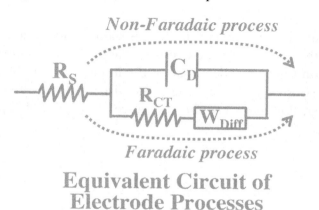

Equivalent Circuit of
Electrode Processes

where $R_S$ is the solution resistance as measured with a conductometer; $C_D$ is the double layer capacitance of the electrode; $R_{CT}$ is the charge transfer resistance related to the Faradaic process; $W_{Diff}$ is the impedance (a Warburg element) originating from the constraint of diffusion.

## Warburg element

A Warburg element (W) is neither a resistance nor a conductance but an imaginary element with the following equivalent circuit:

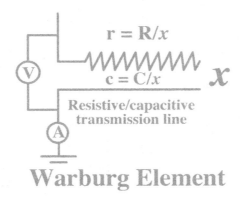

where r and c are respectively the resistance and capacitance per unit length ($x$) of a so-called resistive-capacitive transmission line. At $x = 0$ (on the electrode surface), the impedance will be zero (no diffusion constraint). The longer the distance from the electrode surface ($x$), the larger are the resistance ($R = rx$) and capacitance ($C = cx$).

The element can simulate a diffusion process as follows:

$$I = -\frac{\partial V}{\partial R} = -\frac{1}{r}\frac{\partial V}{\partial x}\text{(Potential gradient)}$$

$$j = -D\frac{\partial Con}{\partial x}\text{(Concentration gradient : Fick's first law)}$$

$$\frac{\partial V}{\partial t} = -\frac{\partial Q}{\partial t}\frac{1}{C} = -\frac{1}{c}\frac{\partial I}{\partial x} = \frac{1}{rc}\frac{\partial^2 V}{\partial x^2}\text{(Potential change)}$$

$$\frac{\partial Con}{\partial t} = D\frac{\partial^2 Con}{\partial x^2}\text{(Concentration change : Fick's second law)}$$

In order to discriminate concentration from capacitance, concentration is symbolized as *Con*. The concentration (*Con*) and flux ( $j$ ) are simulated as the potential ($V$) and current ($I$) at a certain $x$. On the electrode surface ($x = 0$), the potential ($V$) is solved as the semi-integral of current ($I$) as:

$$V(0,t) = -\sqrt{\frac{r}{c}}\frac{d^{-1/2}}{dt^{-1/2}}I(0,t)$$

With a sinusoidal electric perturbation as is usually conducted in an impedance analysis, there will always be a $45°(\pi/4)$ phase lag between $I$ and $V$, just like in a hybrid element of resistance and capacitance.

$$I(t) = I\sin(\omega t + \theta)$$

$$\frac{d^{-1/2}}{d^{-1/2}t}I(t) = -I\frac{1}{\sqrt{\omega}}\sin\left(\omega t + \theta - \frac{\pi}{4}\right)$$

$$V(t) = I\sqrt{\frac{r}{c\omega}}\sin\left(\omega t + \theta - \frac{\pi}{4}\right)$$

An angular speed ($\omega$)-independent quantity, $W$, is introduced to characterize a Warburg element.

$$W = \sqrt{\frac{r}{2c}}$$

The in phase and out of phase impedance ($Z_{in}$ and $Z_{out}$) will be equal in absolute value and inversely related to the square root of the angular speed of the signal.

$$Z_{in} = Z_{out} = \frac{W}{\sqrt{\omega}}$$

The schematics for analyzing the impedance of a resistance and a Warburg element connected in series will be:

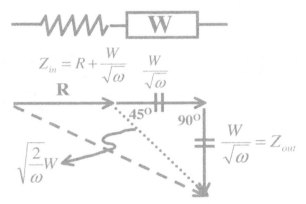

**Impedance "Analysis" of a R/W in Series**

### Impedance of Faradaic pathway

The impedance from the diffusion constraint ($W_{Diff}$) can be expressed as the result of connecting the diffusion constraints of reduced and oxidized species in a serial configuration.

$$W_{Diff} = \frac{RT/nF}{\sqrt{2}nFA}\left(\frac{1}{[R]\sqrt{D_R}} + \frac{1}{[O]\sqrt{D_O}}\right)$$

On the other hand the impedance contributed by the charge transfer process (from ionic current to electronic current or vice versa) on the electrode surface ($R_{CT}$) should be considered in terms of the Butler-Volmer equation.

$$I = nFAj = nFAK_O\{[R]*e^{\alpha_a nF(E-E^{O'})/RT} - [O]*e^{-\alpha_c nF(E-E^{O'})/RT}\}$$

Even with a negligible redox current, a redox process may occur vigorously with the reductive and oxidative processes canceling each other.

$$I_O = nFAK_O[R]*e^{\alpha_a nF(E-E^{O'})/RT} = nFAK_O[O]*e^{-\alpha_c nF(E-E^{O'})/RT} = -I_R$$

The "invisible" current is called the exchange current ($I_{ex}$):

$$I_{ex} = nFAK_O[R]^{\alpha_a}[O]^{\alpha_c}$$

The charge transfer impedance ($R_{CT}$) is the ratio of the normalized potential ($RT/nF$) to the exchange current ($I_{ex}$).

$$R_{CT} = \frac{RT/nF}{I_{ex}}$$

## Impedance analysis of a model electrode process

The analytical result for the combined Faradaic process and non-Faradaic process will be:

$$Z_{in} = R_s + \frac{W + R_{CT}\sqrt{\omega}}{\sqrt{\omega} + 2C_D W\omega + 2C_D^2 W^2 \sqrt{\omega^3} + C_D^2 R_{CT}^2 \sqrt{\omega^5}}$$

$$Z_{out} = -\frac{W + 2C_D W^2 \sqrt{\omega} + 2C_D R_{CT} W\omega + C_D R_{CT}^2 \sqrt{\omega^3}}{\sqrt{\omega} + 2C_D W\omega + 2C_D^2 W^2 \sqrt{\omega^3} + C_D^2 R_{CT}^2 \sqrt{\omega^5}}$$

When the perturbation signal has a low angular speed ($\omega$), the terms with the order of $\omega$ higher than $\omega^{1/2}$ can be neglected:

$$Z_{in} = R_s + R_{CT} + \frac{W}{\sqrt{\omega}}$$

$$Z_{out} = 2C_D W^2 + \frac{W}{\sqrt{\omega}}$$

Both the in phase and out of phase impedance will increase in an equal amount along with the decreasing angular speed ($\omega$). The impedance is under diffusion control, and the plot with different $\omega$ is linear with a slope of nearly 1.

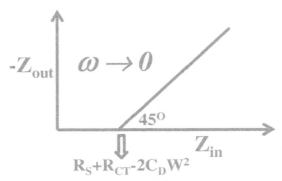

## Impedance of an Electrode Process at Low frequency

The intersect at the abscissa ($Z_{out} = 0$) can be calculated as follows:

$$\frac{W}{\sqrt{\omega}} = -2C_D W^2 (Z_{out} = 0)$$

$$Z_{in} = R_s + R_{CT} - 2C_D W^2$$

As the $\omega$ of the perturbation signal is increased sufficiently:

$$\sqrt{\frac{2}{\omega}}W \ll R_{CT}$$

$$\sqrt{\frac{2}{\omega}}W \ll \frac{1}{\omega C_D}$$

Terms with W can be ignored, and the impedance can be simplified and rearranged as follows:

$$Z_{in} = R_s + \frac{0 + R_{CT}\sqrt{\omega}}{\sqrt{\omega} + 0 + 0 + C_D^2 R_{CT}^2 \sqrt{\omega^5}} = R_s + \frac{R_{CT}}{1 + C_D^2 R_{CT}^2 \omega^2}$$

$$Z_{out} = -\frac{0 + 0 + 0 + C_D R_{CT}^2 \sqrt{\omega^3}}{\sqrt{\omega} + 0 + 0 + C_D^2 R_{CT}^2 \sqrt{\omega^5}} = -\frac{C_D R_{CT}^2 \omega}{1 + C_D^2 R_{CT}^2 \omega^2}$$

$$Z_{out}^2 + \left(Z_{in} - R_s - \frac{R_{CT}}{2}\right)^2 = \left(\frac{R_{CT}}{2}\right)^2$$

The plot will become the following semi-circle:

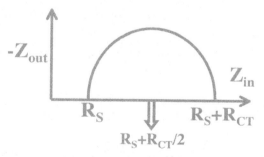

## Impedance of Electrode Process at High frequency

Finally, at infinite frequency ($\omega \to \infty$), the plot will converge at the point ($Z_{in} = R_s$, $Z_{out} = 0$). Using a high frequency signal to suppress

65

the charge transfer and diffusion related impedance ($R_{CT}$ and W), a conductometer can therefore efficiently retrieve the resistance contributed solely by the solution.

By measuring the impedance from low to high frequency, the following impedance plot can be obtained for a typical electrochemical process:

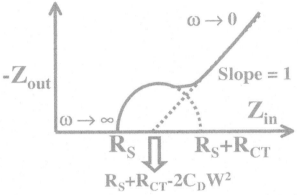

## Impedance of Electrode Process

After analyzing each parameter ($R_S$, $R_{CT}$, $C_D$ and W), the equivalent circuit for the electrode process can be established for further discussion.

## Selection of natural frequency

From the tide to your pulse, the world is full of rhythmic phenomena, but the frequencies of some are too high to be sensed directly, such as light and sound. The following is the electric circuit for a conventional door bell:

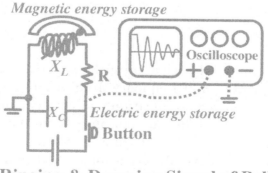

**Ringing & Damping Signal of Bell**

When you press the button, the capacitor is immediately charged with electric energy by the power supply. After releasing the button, the capacitor is now the energy source that activates the inductor and certainly with some energy being dissipated as heat by the inherent resistance of the capacitor and inductor (summarized as R). At this moment, the energy format is converted from electric energy to its magnetic counterpart, and the bell starts its first strike from the induced magnetic field.

The magnetic field will decay along with the decaying discharge current from the capacitor, which will induce a voltage across the inductor to charge the capacitor again. This time, the energy will flow from the inductor (the magnetic energy storage) to the electric counterpart. The story will return to the very beginning — the release of the button, and the resulting oscillating energy transfer will trigger a rhythmic striking of the bell with a frequency defined as follows:

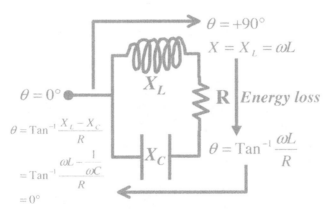

# Frequency Determining Network

The phase shift of the signal through a capacitor or inductor is strongly correlated with its frequency. To deal with the phase shift through these elements, we need only to consider the imaginary parts, the reactances ($X_C$ for the capacitor; $X_L$ for the inductor) of their impedances ($Z_C$ for the capacitor; $Z_L$ for the inductor). The real parts, the resistances of the capacitor and inductor, are merged as R to ease the analysis.

At the moment the button is released, the signal is actually composed of signals of various frequencies, but finally it is the signals with zero phase shift $(0, 2\pi, 4\pi, \ldots\ldots)$ after a whole cycle that will "survive."

$$\omega L = \frac{1}{\omega C}$$

$$\omega = \frac{1}{\sqrt{LC}}$$

$$f = \frac{\omega}{2\pi} = \frac{1}{2\pi\sqrt{LC}}$$

The above frequency is called the "natural frequency" of the LC oscillator.

### Barkhausen criteria

Unfortunately, the oscillation phenomenon will be damped by the energy dissipated on the real parts of the impedances, R. A damping factor is therefore defined as:

$$D = \frac{R}{X_L} = \frac{R}{X_C} = \frac{\text{Energy dissipation}}{\text{Tendency to vibrate}} = \frac{1}{Q}$$

For a circuit with large energy storage ($X_L$ or $X_C$) and small energy dissipation (R), the damping will be negligible and the oscillation circuit is called a high Q (quality value) oscillator.

For a sustained oscillation, the decaying signal should be amplified to its original amplitude.

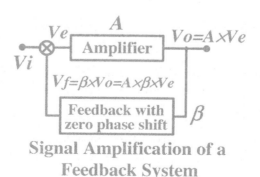

**Signal Amplification of a Feedback System**

With the above scheme, the gain of the signal will be:

$$Gain = \frac{V_o}{V_i} = \frac{V_o}{V_e - V_f} = \frac{A \times V_e}{V_e - A \times \beta \times V_e} = \frac{A}{1 - A \times \beta}$$

An oscillator is actually a looped signal with zero input, so the gain at the oscillating frequency should be infinite. The denominator of the right-handed term should be zero at the frequency as defined by the following Barkhausen Criteria:

$$A(S) \times \beta(S) = 1\angle 0° = 1 + 0j \, (\text{Complex number})$$
$$S = j\omega$$

## Colpitts oscillator

Based on the Barkhausen Criteria, the following Colpitts oscillator can be analyzed with respect to its imaginary and then its real part:

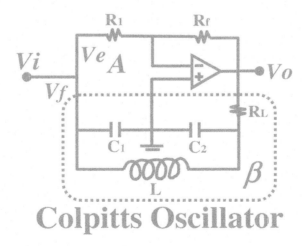

At first, the oscillation frequency is determined by the (zero) phase shift of the frequency determining network (the dashed box).

$$\omega_o = \sqrt{\frac{C_1 + C_2}{C_1 C_2 L}} \, (\text{from the imaginary part})$$

Of course, the natural frequency $(\omega_o)$ can be tuned by adjusting the capacitances $(C_1$ and $C_2)$ and inductance $(L)$. Substituting $\omega = \omega_o$ into the real part gives:

$$\frac{R_f}{R_L} = \frac{C_2}{C_1} \text{(from the real part)}$$

Finally, the DC gain (G) should be set by adjusting $R_1$ to sustain the oscillation.

$$G = \frac{R_f}{R_1} \text{(Gain of the inverting amplifier unit)}$$

A quartz crystal oscillator can be made by replacing the inductor in a Colpitts oscillator with a piezoelectric quartz crystal.

### Quartz crystal oscillator

In an analysis of impedance with an impedance analyzer, a quartz crystal will show the following reactance which can be simulated by several elements:

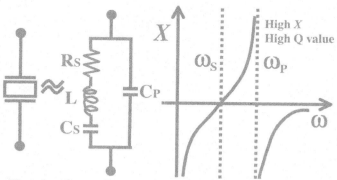

**Equivalent Circuit and Reactance of an AT-Cut Quartz Crystal**

For an AT-cut, 2 MHz quartz crystal, the parameters for the elements in the above equivalent circuit have the following typical values: $R_S = 82\Omega$; $L = 0.52H$; $C_S = 0.0122pF$; $C_P = 4.27pF$; $C_P/C_S = 350$. Neglecting $R_S$, the equation for the impedance will be:

$$Z(S) = \cfrac{1}{\cfrac{1}{SC_P} + \cfrac{1}{SL + \cfrac{1}{SC_S}}} = \cfrac{1}{SC_P} \bullet \cfrac{S^2 + \cfrac{1}{LC_S}}{S^2 + \cfrac{1}{L\left(\cfrac{C_SC_P}{C_P + C_S}\right)}}$$

$$Z(j\omega) = -\left(\frac{j}{\omega C_P}\right)\left(\frac{\omega^2 - \omega_S{}^2}{\omega^2 - \omega_P{}^2}\right)$$

$$\omega_S = \frac{1}{\sqrt{LC_S}}$$

$$\omega_P = \sqrt{\frac{C_S + C_P}{C_SC_PC_L}} \cong \frac{1}{\sqrt{LC_S}} (\because C_P \gg C_S)$$

To serve as the inductor in a Colpitts oscillator, the reactance should be positive, which is only possible in the range of $\omega_S < \omega < \omega_P$. Since $\omega_S$ is very close to $\omega_P$, the oscillator will have a very stable natural frequency.

$$f = \frac{\omega}{2\pi} = \frac{1}{2\pi\sqrt{LC_S}} = \frac{1}{2\pi\sqrt{0.52H \times 0.0122pF}} = 1989436Hz \cong 2MHz$$

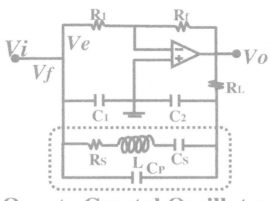

## Quartz Crystal Oscillator

## Quartz crystal microbalance (QCM)

With its extremely stable natural frequency, a quartz crystal oscillator is often used as a clock and for timing in digital circuits. However, the natural frequency ($f_o$) will slightly decrease ($\Delta f$) when the oscillation is damped by the adsorption of a tiny mass ($\Delta m$) on the quartz surface.

$$\Delta f = -\frac{2nf_o^2}{\sqrt{\mu\rho}} \times \frac{\Delta m}{A}$$

In the above Sauebrey equation, $A$ is the surface area of an AT-cut quartz crystal; $\rho$ is the density of the crystal with the value of 2.65 gcm$^{-3}$; $\mu$ is the shear modulus of the crystal ($2.95 \times 10^{11}$ gcm-1S$^{-2}$); $n$ is the harmonic number (or overtone) of the oscillation. The oscillation frequency will be the natural frequency multiplied by $n$ (e.g. $n = 1$ for a 2 MHz crystal oscillating at 2 MHz).

From the shift in oscillation frequency, it is possible to detect nanogram order per cm$^2$ of mass change on the quartz surface. The highly sensitive mass sensor (balance) is particularly useful in investigating bio-affinity phenomena such as the binding of antigen and antibody in a dynamic (time-resolved) manner, which can hardly be achieved by conventional binding assays.

Moreover, a redox process on an electrode mounted on the quartz surface can be monitored from the Faradaic current. Simultaneously with electrochemical detection, the electrode process can also be traced by the aforementioned mass dependent frequency shift. The coupled technique is called EQCM (electrochemical quartz microbalance), a promising multi-mode tool for surface chemistry.

# Electrophoresis

## The basic separation technologies for modern bioscience

Modern biotechniques often deal with proteins, DNA and other nano-scaled biomolecules. In the microscopic world with high charge to mass ratios (Q/M), electrostatic force is much more important than gravitational force. Biomolecules can therefore be easily separated under a suitable electric field. As a consequence, electrophoresis has become the major technology for purifying these nano-particles for further investigation or applications.

## Gel electrophoresis

However, the motion driven by an electric field (migration) is continuously disturbed by the convection of the medium and the diffusion caused by the Brownian motion of the molecules. To enhance the difference in migration speed (i.e. to prevent the blurring of the bands in the gel), the other motions should be suppressed. Gel electrophoresis uses the technique of restricting convection and diffusion in a gel such as agar during electrophoresis. The principles of this efficient technique can be easily understood and will be omitted for brevity. However, after reading the electrochemical concepts in the previous chapters, naturally, the readers will ask the question that follows.

### Did redox reactions occur during electrophoresis?

For gel electrophoresis, some hundreds of volts are imposed on the electrodes; redox reactions are certainly inevitable. However, for an electrophoresis system, the solution resistance ($Z_s$) is usually very high and will dominate practically all of the electric work. The energy is therefore used almost entirely for separation.

As seen from the bubbling on the platinum wire connected to the "negative" pole of a gel electrophoresis system, redox reactions must occur to evolve hydrogen gas. Consequently, noble metals such as platinum should be used to avoid possible contamination during redox processes.

### Capillary electrophoresis

Based on the theoretical plate theory for chromatography, we need only to elongate the electrophoretic path to increase the separation efficiency. However, it is inconvenient to extend a whole system in a linear manner. In capillary electrophoresis (CE), the electrophoretic path is lead by a curled capillary filled with the electrolytes for electrophoresis (carrier solution). The capillary is made of glass or fused silica of about 50 $\mu$m i.d. and some decimeters in length. After introducing several nano-liters of sample into the capillary, tens of kV is imposed to start the electrophoresis process.

### Electroosmotic flow

Filled with carrier solution with a pH higher than about 3, the inner surface of the wall of a fused silica capillary will possess negative charges. To maintain electroneutrality, there must be "counter" cations to surround the negatively charged surface. In other words, near the wall, there will be a deviation from electroneutrality with a positive ion atmosphere. The negative charge density generated by the "stationary" negative charges on the wall will be "damped" by the mobile positive charges. Considering also the Brownian motion, the profile of the charge density will become:

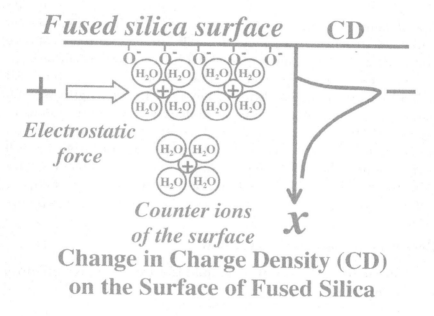

**Change in Charge Density (CD) on the Surface of Fused Silica**

Consequently, there is an excess of "mobile" cations near the wall, which cannot be easily disturbed by thermal motion in an environment with low Reynolds number. Under the imposed electric field, the cations near the wall along with the hydrating water molecules will be dragged toward the negative pole. Due to the inherent viscosity of water, or more precisely the intermolecular hydrogen bonds, all the electrolytes and water molecules will be dragged simultaneously with a velocity ($v_{eo}$) described by the following Smoluchowski equation:

$$v_{eo}(x) = -\frac{\varepsilon}{\eta} E[\zeta - \phi(x)]$$

Where $v_{eo}$ is the velocity of the electroosmotic flow with the dimensions of mS$^{-1}$, not m$^3$S$^{-1}$ for a volumetric flow rate; $\varepsilon$ is the permittivity of the carrier solution which is the product of the dielectric constant ($\varepsilon_r$) and the permittivity in vacuum ($\varepsilon_o$); $\eta$ is the viscosity of the carrier solution; $E$ is the external electric field strength, the ratio of imposed voltage ($V$) to the capillary length ($L$) for a parallel field; $\zeta$ is the zeta potential generated by the negative charge density on the capillary wall, for a fused silica capillary, the typical value is −75mV; $\varphi$ is the local potential determined by the stationary negative charges on the wall and the counter cationic atmosphere, which is a function of the distance from wall surface ($x$).

The local potential will decrease from the zeta potential (−75mV; $x = 0$) immediately to zero in a few Debye lengths (several nanometers). Therefore, across a capillary with an inner radius of typically 50000 nm, the velocity holds constant except in the region very close to the wall surface where the frictional force from the wall surface retards the flow. The dragging by the phenomena of surface charge excess causes the profile of electroosmotic flow to be quite different from a profile that is driven by a pressure drop using a pump.

## Between plug flow and laminar flow

Lamina flow occurs when the fluid is driven by a pressure drop such as in HPLC.

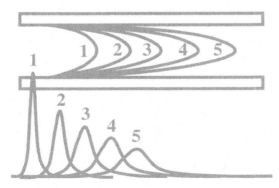

**Laminar Flow and the Peak Profiles
at Different Time and Distance**

In HPLC, the flow close to the wall is retarded by the frictional force, and the peak profile (the signal for analysis) will be blurred along with the retention time. On the other hand, for an electroosmotic flow, the fluid is pulled by the charge imbalance on the wall surface; the effect of frictional force is counteracted to obtain the following plug flow with a sharper peak profile:

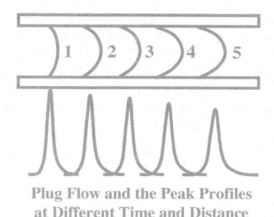

**Plug Flow and the Peak Profiles
at Different Time and Distance**

As a consequence, the electroosmosis-based CE technique has outstanding separation efficiency (high theoretical plate number) and becomes the basis of automatic DNA sequencing for the human genome project. However, to develop an optimized CE method, care must be taken with several important parameters including the ionic strength and temperature of the carrier solution.

### Effects of ionic strength and temperature

As the ionic strength increases, the Debye radius will decrease with the flow profile approximating ideal plug flow. Moreover, due to the increased shielding effect, unnecessary electrostatic adsorption on the capillary surface is suppressed and the electroosmotic flow driven by surface phenomena will improve in reproducibility. However, the increased conductivity of the carrier solution lowers the resistance of the capillary and results in joule heating problems.

The temperature of the carrier solution, especially in the center of the capillary, will rise if the resistive heating problem is not properly controlled. The rise in temperature will lower the solution viscosity and resistivity (the inverse of conductivity), and a vicious spiral of — uncontrolled temperature increase $\rightarrow$ conductivity increase $\rightarrow$ joule heating problems — start. Additionally, a temperature and thus viscosity gradient will build-up from the wall surface (low temperature & high viscosity) to the center (high temperature & low viscosity) of the capillary, which will disturb the plug flow and lead to an undesirable laminar one.

Therefore, prior to an experiment, the author used to determine the limiting voltage $(V_{limit})$ for a CE system by measuring the characteristic I/V curve. The curve will deviate from ohmic behavior $(I/V = R = \text{constant})$ at the limiting voltage.

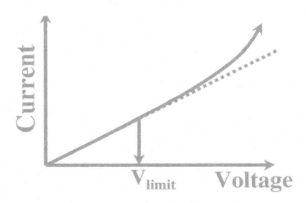

## I/V Curve for Determination of Limiting Voltage ($V_{limit}$) for Capillary Electrophoresis

With the separation voltage lower than $V_{limit}$, the electrophoresis will be well-controlled in temperature and flow profile. As another drawback of high ionic strength, the zeta potential is prone to decrease with the increase in ionic strength, which will result in somewhat of a decrease in electroosmotic flow and the analysis speed.

# Appendices

# Misconceptions in High School Electrochemistry

The following reaction is probably the most familiar from the electrochemistry lectures of our high schools:

$$Zn_{(S)} + Cu^{2+}{}_{(aq)} \rightarrow Zn^{2+}{}_{(aq)} + Cu_{(S)}$$

However, after putting water, a zinc plate and cuprous sulfate ($CuSO_4$) into a beaker, no reaction will occur as shown in the left-hand side of the following figure:

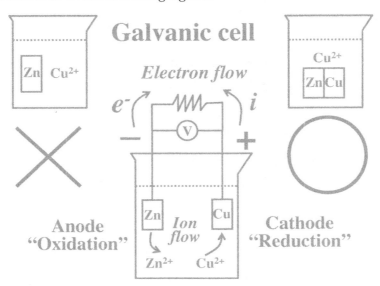

Is there something wrong with our textbooks? Next, we tie the zinc plate to a copper plate and put them into the beaker again as shown in the right hand side of the above figure. The "textbook" reaction does occur with the zinc plate getting thinner and the copper plate getting thicker.

In reality, an electrochemical reaction cannot occur independently without electron transfer, so the real story (the mechanism) should be:

$$Zn_{(S)} \rightarrow Zn^{2+} + 2e^-$$

$$Cu^{2+} + 2e^- \rightarrow Cu_{(S)}$$

Since the electrons generated on the surface of the zinc plate cannot straightforwardly enter the water phase, we should make a "path of escape" for them. Instead of being in direct contact with a copper plate (the right-hand beaker in the above), the zinc plate can also be guided through a wire to conduct its electrons to the copper (the middle beaker in the above). In this manner, a voltage (about 1.5V Cu/Zn) can be measured between the two plates with the positive pole (Cu) being the cathode for the reduction process, exactly as a galvanic cell does.

# Laplace & Fourier Transformation

### Solving differential equations by algebraic equations

For a large exponential number such as $228^{28}$, it would be convenient to approximate its true value by inverting through its logarithmic value.

$x = 228^{28} = 228 \times 228 \times 228\ldots\ldots$

$\log x = \log 228 + \log 228 + \log 228\ldots.. = 28 \log 228 \cong 66.02 = 66 + 0.02$

$x = 10^{66} \times 10^{0.02} \cong 1.05 \times 10^{66}$

A burdensome multiplication is first converted into a different "world" using only simple addition, and then the result of the addition is inversely converted into an approximate value.

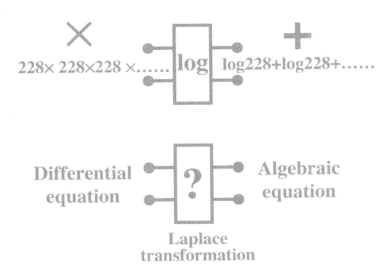

Laplace transformation is another magic box that converts between differential and algebraic equations.

## Curing cancer by solving differential equations

After intravenous injection of an anti-cancer agent, the plasma (blood) concentration of the drug will immediately rise to certain initial concentration, $C_{in}$, and then decline by the elimination and detoxification effects of the kidney and liver. Assuming first order kinetics, the rate of decline (R) will be proportional to the concentration (C) itself.

$$R = -\frac{dC}{dt} = K_{out}C$$

Where $K_{out}$ is the rate constant with dimensions of the inverse of time. The differential equation can be easily solved by separation of variables C and t:

$$-\frac{dC}{C} = K_{out}dt$$

$$\ln C \Big[\begin{smallmatrix} t \\ t=0 \end{smallmatrix} = -K_{out}t \Big[\begin{smallmatrix} t \\ t=0 \end{smallmatrix}$$

$$\ln C - \ln C_{in} = -K_{out}t$$

$$C = C_{in}e^{-K_{out}t} = C_{in}e^{-t/\tau_{out}}$$

The concentration is predicted to decline exponentially from $C_{in}$ to zero with a time constant of $\tau_{out}$, the inverse of $K_{out}$.

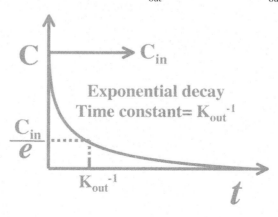

If the drug is administrated by infusion, a term of constant input $(K_{in})$ should be added in the differential equation.

$$\frac{dC}{dt} = K_{in} - K_{out}C$$

This time, the variables cannot be separated straight-forwardly, so the concentration $(C)$ equation (in the time domain) is to be transformed with respect to $t$ into the Laplace domain (S domain).

$$\frac{dC}{dt} = K_{in} - K_{out}C$$

Laplace transformation

$$SF(S) - C(0) = K_{in} \times \frac{1}{S} - K_{out} \times F(S)$$

Substituting the initial concentration $c(0) = 0$ to the above transform, the "algebraic equation" can be solved for the concentration in the S domain, $F(S)$. After that, the algebraic equation is rearranged into a format that can be easily transformed back to the time domain.

$$SF(S) = \frac{K_{in}}{S} - K_{out}F(S)$$

$$(S + K_{out})F(S) = \frac{K_{in}}{S}$$

$$F(S) = \frac{K_{in}}{S(S + K_{out})} = \frac{K_{in}}{K_{out}}\left[\frac{1}{S} - \frac{1}{S + K_{out}}\right]$$

The inverse transformation is as follows:

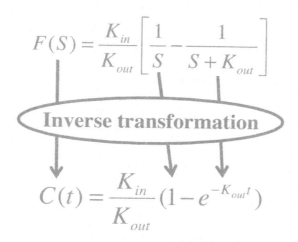

$$F(S) = \frac{K_{in}}{K_{out}}\left[\frac{1}{S} - \frac{1}{S + K_{out}}\right]$$

**Inverse transformation**

$$C(t) = \frac{K_{in}}{K_{out}}(1 - e^{-K_{out}t})$$

Without using any skill from calculus, the differential equation is solved in a simply algebraic manner. The plasma concentration will rise exponentially from zero to the concentration $K_{in}/K_{out}$.

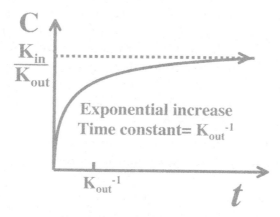

Due to the inherent toxicity of the anti-cancer agent, the plasma concentration should be carefully controlled between the effective concentration ($C_{effective}$) and the toxic concentration ($C_{toxic}$) as the following:

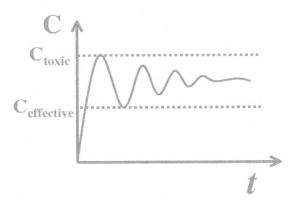

Knowing the kinetics of drug metabolism and the methods and timing of drug administration is pivotal for modern cancer therapeutics.

## Laplace transformation

A Laplace transformation with respect to time ($t$) is used to integrate the product of an objective time-relating function, $f(t)$ and $e^{St}$ from t=0 to infinity. Due to the converging effect of $e^{St}$, the integral will converge into a function of $S$, $F(S)$.

$$L\{f(t)\} = \int_0^\infty f(t)e^{-St}dt = F(S)$$

The function is said to be transformed from the time domain to the S domain (or Laplace domain) where it can be dealt with algebraically.

The Laplace transforms and the inverse transforms of some simple functions can be obtained directly by integration such as:

$$f(t) = 1$$

$$L\{1\} = \int_0^\infty e^{-St}dt = [-\frac{1}{S}e^{-St}]_0^\infty = \frac{1}{S}$$

and

$$f(t) = e^{-\alpha t}$$

$$L(e^{-\alpha t}) = \int_0^\infty e^{-\alpha t} e^{-St} dt = \int_0^\infty e^{-(S+\alpha)t} dt = [-\frac{1}{S+\alpha} e^{-(S+\alpha)t} ]_0^\infty = \frac{1}{S+\alpha}$$

Some others are very complicated, and it is better to consult the table in the back of a mathematics textbook. The following are those most frequently used in biomedical areas:

$$f(t) \xrightarrow{\text{Laplace transformation}} F(S)$$

| $f(t)$ | $F(S)$ |
|---|---|
| $1$ | $\dfrac{1}{S}$ |
| $t^n$ | $\dfrac{n!}{S^{n+1}}$ |
| $e^{\alpha t}$ | $\dfrac{1}{S-\alpha}$ |
| $f'(t)$ | $SF(S) - f(0)$ |
| $f''(t)$ | $S^2 F(S) - Sf(0) - f'(0)$ |

## From Laplace to Fourier transformation

The physical meaning of the transformed function in the S domain is somewhat hard to imagine, but for a Fourier transformation, S represents the "imaginary" frequency ($S = j\omega$), the inverse quantity of time.

$$S = j\omega$$

$$F(j\omega) = \int_0^\infty f(t) e^{-j\omega t} dt$$

The Fourier transformation is therefore a special case of the Laplace transformation, and a time-dependent function will be transformed into a function of frequency, just like the "histogram" of the equalizer of your stereo.

$$H(j\omega) = \int_{-\infty}^{\infty} h(t)e^{-j\omega t}dt$$

After introducing the Euler equation into the exponential term, $e^{j\omega t}$, the time-dependent signal $h(t)$ such as a sound (the vibration of air) will be converted to a function of frequency $H(j\omega)$ that can be subsequently separated into a real and imaginary part.

$$e^{j\theta} = \cos(\theta) + j\sin(\theta) : \text{Euler equation}$$
$$\theta = -\omega t$$

$$H(j\omega) = \int_{-\infty}^{\infty} h(t)e^{-j\omega t}dt = \int_{-\infty}^{\infty} h(t)[\cos(-\omega t) + j\sin(-\omega t)]dt$$

$$= \int_{-\infty}^{\infty} Amplitude(t)Phasor(-\omega t)dt$$

$$= \int_{-\infty}^{\infty} h(t)\cos(-\omega t)dt + j\int_{-\infty}^{\infty} h(t)\sin(-\omega t)dt$$

The histogram of the equalizer is actually the summation of the real part of the transforms at different frequencies, a "composition" of sinusoidal signals with different frequencies.

$$Histogram = \sum H(j\omega)$$

## Versatility of the Fourier transformation in signal processing

To retrieve or remove the contribution of signals with a certain frequency, a time dependent signal can first be transformed into the frequency domain. After retrieving or removing the target contribution from "the histogram", the processed histogram is then inversely transformed to the time domain. Certain desired signals can be selectively retrieved, or the noise can be removed artificially in a digital manner.

The characteristic histogram of your sound or other time-dependent signals can serve as a personal pattern or fingerprint of your signals.

One can also convert a pulse signal with a short duration to "synthesize" a signal with a wide spectrum of frequency. This has become a basic technique in spectroscopy such as IR and NMR.

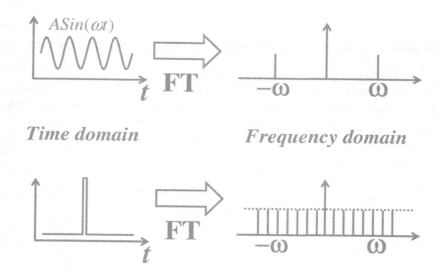

*Time domain*          *Frequency domain*

## Basic Electric Concepts

### Four-terminal networks: the output as the function of an input

Most electric circuits are designed as a device to convert an input signal to a defined output signal as in the scheme (E) in the following figure. For example, in the same figure, circuit (A) with a variable resistor is for reducing the voltage to a defined fraction as in (B):

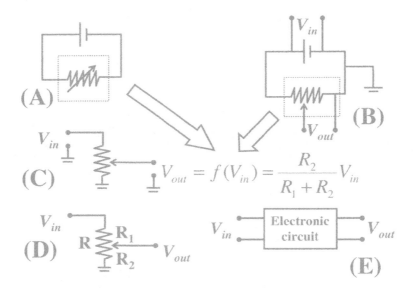

$$V_{out} = f(V_{in}) = \frac{R_2}{R_1 + R_2} V_{in}$$

As modern electric circuits become more and more complicated, this kind of circuit is generally simplified to (C) and finally (D) for brevity. In this case, both the negative terminals of $V_{in}$ and $V_{out}$ are connected to a common ground. Since the input impedance of $V_{out}$ is large, the current flowing into the terminal is negligible.

$$I = \frac{V_{out}}{R_2} = \frac{V_{in}}{R_1 + R_2}$$

The relationship of the input and output voltage signal will be:

$$V_{out} = \frac{R_2}{R_1 + R_2} V_{in}$$

## Charging a capacitor

It becomes more and more difficult to completely charge a capacitor with a constant voltage ($V$), and the current ($I$) will decay as a consequence. The charge ($Q$) accumulated on the capacitor will result in a capacitor voltage ($V_C$) proportional to a constant ($C$). The constant ($C$) is defined as the capacitance of the capacitor with the SI unit of farads, F.

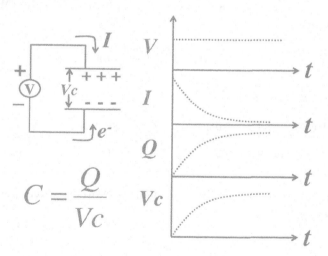

The charging current is a function of time as the following Ohm's law for a capacitor:

$$I(t) = \frac{dQ}{dt} = C \frac{dV_C}{dt}$$

## Charging a capacitor with an A.C. voltage

It seems to be meaningless to charge a capacitor with an A.C. voltage source ($V$); however, the phase of the charging current will be delayed by $\pi/2$. Why?

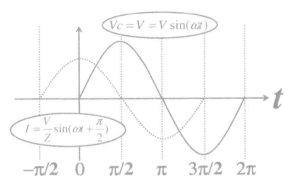

The reason for the 90° phase lag can be resolved by Ohm's law for a capacitor.

$$I(t) = C\frac{dV_C}{dt} \approx C\frac{dV(t)}{dt} = C\frac{dV\sin(\omega t)}{dt} = \omega CV\cos(\omega t)$$

$$= \omega CV\sin\left(\omega t + \frac{\pi}{2}\right)$$

$$Z = \frac{V(t)}{I(t)} = \frac{V\sin(\omega t)}{\omega CV\sin\left(\omega t + \frac{\pi}{2}\right)} = \frac{1}{\omega C}\times\frac{\sin(\omega t)}{\sin\left(\omega t + \frac{\pi}{2}\right)} = \frac{1}{\omega C}\angle -90°$$

Regardless of phase, the size of impedance is inversely related to frequency.

$$Z = \frac{1}{\omega C}$$

**Frequency domain**

Considering the phase shift, the expression in polar form is more obvious in showing the exact value of the phase angle, but the rectangular form using a complex number will be more convenient for mathematical analysis. The "A.C. resistance" included with the phase shift is called impedance.

### Analyzing impedance with a complex number

Using the Euler equation, the polar form of a quantity with phase angle ($\theta$) such as an impedance Z can be resolved into a real and imaginary part as follows:

$$e^{j\theta} = \cos(\theta) + j\sin(\theta)$$

$$Z = |Z|\angle\theta = |Z|e^{j\theta} = |Z|\cos(\theta) + j|Z|\sin(\theta)$$

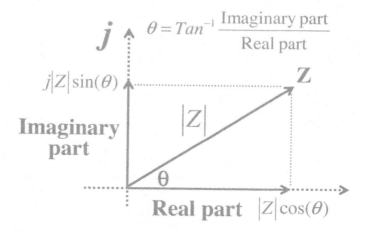

The impedance of a capacitor can be converted to a complex function of frequency.

$$Zc = \frac{V(t)}{I(t)} = \frac{1}{\omega C}\angle-90° = \frac{1}{\omega C}\left[\cos\left(\frac{-\pi}{2}\right) + j\sin\left(\frac{-\pi}{2}\right)\right] = \frac{-j}{\omega C} = \frac{1}{j\omega C}$$

Using $S = j\omega$ as in Fourier transformation, the impedance will become:

$$Zc = \frac{1}{j\omega C} = \frac{1}{SC} = \frac{1}{\omega C} \angle -90°$$

With similar treatments based on Ohm's law for an inductor, the impedance ($Z_L$) will be related to the inductance ($L$) with a positive phase shift of $\pi/2$.

$$Z_L = j\omega L = SL = \omega L \angle 90°$$

## Low pass filter

Most of the noises in an electric circuit are in the high frequency region, which can be removed or reduced by the following low pass filter:

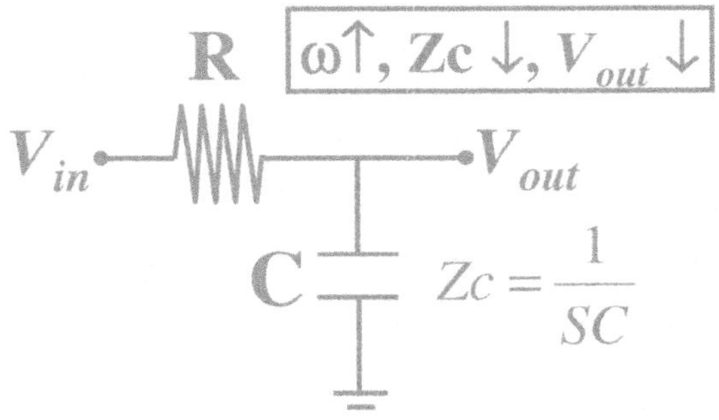

**1-Order low-pass filter**

Analyzing by complex number, the gain ($G = V_{out}/V_{in}$) of the circuit will be a function of frequency.

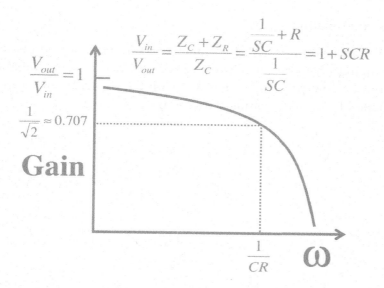

The product of the capacitance (C) of the capacitor and the resistance (R) of the resistor is called the time constant of the filter, and the low pass filter will have a cut-off frequency at the inverse of CR ($\omega_{cutoff}$ = 1/CR). For a signal at the cut-off frequency, the gain (G) will be:

$$S = j\omega_{cutoff} = j / CR$$

$$G = \left|\frac{1}{1+SCR}\right| = \left|\frac{1}{1+j}\right| = \frac{1}{\sqrt{2}} \cong 0.707$$

The phase shift can be calculated to be -45° as follows:

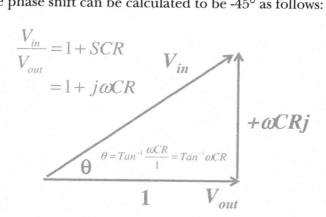

## Filter combination

A high pass filter with the same time constant can be made by exchanging the position of the resistor and capacitor in the above mentioned low pass filter. The gain and the phase shift can be calculated by similar treatment as the low pass filter.

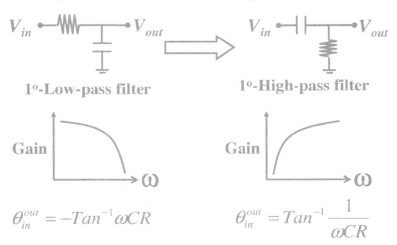

**1°-Low-pass filter**

**1°-High-pass filter**

$$\theta_{in}^{out} = -Tan^{-1}\omega CR$$

$$\theta_{in}^{out} = Tan^{-1}\frac{1}{\omega CR}$$

By connecting three equivalent low pass filters in series, a high performance filter (or a 3$^{rd}$ order filter) with a sharper signal cut-off profile can be obtained.

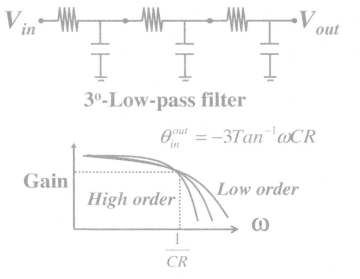

**3°-Low-pass filter**

$$\theta_{in}^{out} = -3Tan^{-1}\omega CR$$

The time constant will not change, and the phase shift will be three times as large as the first order filter.

Next, connecting a low and high pass filter in series creates a band pass filter allowing only the passage of signals with a frequency around the inverse of the time constant $(1/CR)$.

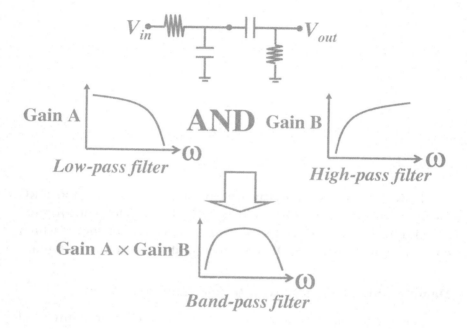

The signal will pass through a filter-permitting low frequency "and" then another allowing only high frequency; it is the logic operator "AND" that restricts the passage of both the low and high frequency signals. The band pass behavior can also be predicted by multiplying the gains of both filters.

On the other hand, the following parallel connection will lead to a band rejection filter. The parallel configuration is just like a logic operator "OR", either high "or" low frequency signals can pass. The signals with frequency around $1/CR$ will be rejected. A band rejection filter with a time constant of $1/60Hz$ can be used to remove the noise from the 60 Hz family A.C. power source.

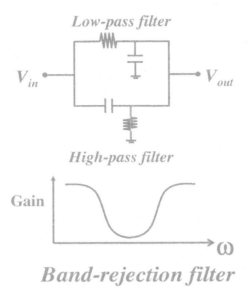

*Low-pass filter*

$V_{in}$      $V_{out}$

*High-pass filter*

Gain

$\omega$

## *Band-rejection filter*

However, signals will decay after passing through the above RC (resistor and capacitor)-based filters. To bring back the original amplitude, it is necessary to use active elements (elements which consume energy) such as OP-amps to amplify the attenuated signals.

### Dealing with analog signals with operational amplifiers

An operational amplifier, or OPA, is the most frequently used integrated circuit (IC) for handling analog signals. An ideal OPA has

## Operational Amplifier

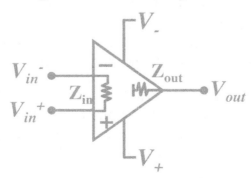

$V_-$

$V_{in}^-$   $Z_{out}$

$Z_{in}$   $V_{out}$

$V_{in}^+$

$V_+$

the following three peculiarities:
   1. Infinite input impedance ($Z_{in} \rightarrow \infty$)
   2. Zero output impedance ($Z_{out} \rightarrow 0$)
   3. Infinite differential amplification $\quad Gain = \dfrac{V_{out}}{V_{in}{}^+ - V_{in}{}^-} \rightarrow \infty$

The first and second properties are obviously for impedance matching during the transmission of weak signals. The third property is nothing but the definition of a comparator.

# Voltage Comparator

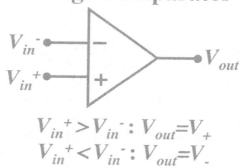

$$V_{in}{}^+ > V_{in}{}^- : V_{out} = V_+$$
$$V_{in}{}^+ < V_{in}{}^- : V_{out} = V_-$$

With $V_{in}{}^+$ slightly higher than $V_{in}{}^-$, the $V_{out}$ will be infinity. In reality, the voltage will be limited by the positive terminal ($V^+$) of the power source supplying the IC. This is usually about 12V for a conventional OPA. Certainly the negative output will be restricted by the negative terminal ($V^-$) of the power source. In this configuration, the IC behaves more like a circuit designed for handling digital signals.

## Voltage follower

In the next, putting a jumper between $V_{in}{}^-$ and $V_{out}$ to make a feedback bypass causes the circuit to become a voltage follower.

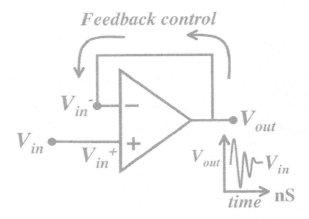

Even if a tiny difference exists between $V_{in}^+$ and $V_{in}^-$, $V_{out}$ will be enormously amplified by the infinite differential gain, the third ideal property of an OPA. The signal will be fed back to $V_{in}^-$ to suppress the voltage difference within nanoseconds. As a consequence, $V_{in}^+$ will be practically equivalent to $V_{in}^-$. This is called a virtual short between the two input terminals of an OPA with negative feedback. Surely we cannot directly observe the extremely fast negative feedback control process.

From a different but straightforward viewpoint, $V_{in}^+$ must be equal to $V_{in}^-$ to achieve the infinite differential gain.

$$\because Gain = \frac{V_{out}}{V_{in}^+ - V_{in}^-} \rightarrow \infty$$

$$\therefore V_{in}^+ = V_{in}^-$$

Finally, since $V_{in}^-$ is connected directly to $V_{out}$, $V_{in}^-$ will be equal to $V_{in}^+$ and then $V_{out}$.

# Voltage Follower

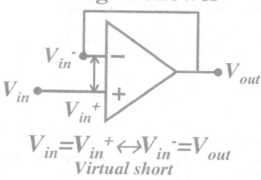

$$V_{in}=V_{in}^+\leftrightarrow V_{in}^-=V_{out}$$
*Virtual short*

A voltage follower can be used to avoid the problem of impedance matching.

### Inverting amplifier

The following is a circuit with a defined power of amplification:

# Inverting Amplifier

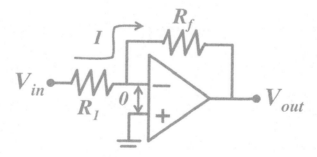

The potential at the negative (or inverting) input terminal ($V_{in}^-$) of the OPA will be virtually grounded as a result of the virtual short between the two input terminals. Due to the high input impedance of an OPA, practically all current will flow through the feedback bypass with the following ohmic relationships:

$$I = \frac{0 - V_{in}}{R_1} = \frac{V_{out} - 0}{R_f}$$

$$V_{out} = -\frac{R_f}{R_1} V_{in}$$

Consequently, the gain will be a negative value defined by the ratio of two resistors $(-R_f / R_1)$. If the source signal is an A.C. signal, the phase will be inverted with a 180° shift.

### From active filter to oscillator

The attenuated signal $(V_x)$ from a passive filter (e.g. RC filter) can be fed to the non-inverting terminal (the positive input terminal) of an OPA for amplification without phase inversion. The passive filter is thus upgraded to an active filter.

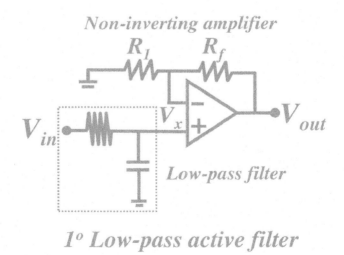

*1° Low-pass active filter*

The transfer function, $T(S)$, of the active filter can be calculated as follows:

$$\frac{V_x}{V_{in}} = \frac{Zc}{Zc+R} = \frac{1}{1+SRC}$$

$$T(S) = \frac{V_{out}}{V_{in}} = \frac{V_{out}}{V_x} \times \frac{V_x}{V_{in}} = \frac{R_f+R_1}{R_1} \times \frac{1}{1+SRC} = (1+\frac{R_f}{R_1}) \times \frac{1}{1+SRC}$$

The transfer function contains the DC gain $(1+R_f/R_1)$ defined by the non-inverting amplifier and the frequency-dependent filter effect $(1 / 1+SRC)$ along with the phase shift.

In the next circuit, let's see what will happen if the output $(V_{out})$ of the following active filter is fed back to the input $(V_{in})$.

## *Inverting amplifier*

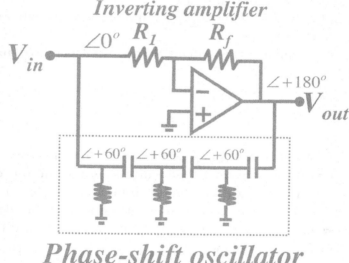

## *Phase-shift oscillator*

The input signal may be composed of signals with various frequencies, but these signals will interfere with each other until a signal with a certain frequency survives. That will be the signal with a zero phase shift after the feedback. The frequency of the surviving signal is called the oscillation frequency of the phase-shift oscillator. With the active element, the OPA, to amplify the attenuated feedback signal, the oscillating signal can persist practically without an input signal. From a different viewpoint, an oscillation circuit can be considered to be a high performance band-pass filter.

# Reynolds Number

Observing the way a paramecium does its "swirling swimming" under a microscope, you can figure out that the tiny creature actually has a hard time conquering the viscosity of water—just like a human being swimming in syrup. Let's take a different look at the "American football" style of swirling swimming through an analysis of the Reynolds number.

Reynolds number (R) is actually a dimensionless ratio of inertial force ($F_I$) to viscous force ($F_V$):

$$R = \frac{F_I}{F_V}$$

For an object with a mass $m$, density $D$ and effective length $L$ swimming at a speed $v$ in a fluid with viscosity $\eta$, the speed of a molecule in the fluid will accelerate from zero (at the moment the object touches the molecule) to $v$ (at the moment it takes the molecule swimming together with it by frictional force) in a duration of $L/v$. The acceleration $a$ will be:

$$a = \frac{dv}{dt} \approx \frac{v-0}{L/v} = \frac{v^2}{L}$$

The inertial force ($F_I$) will be:

$$F_I = m \times a \approx (DL^3) \times \left(\frac{v^2}{L}\right) = Dv^2 L^2$$

Viscous stress $S$ is the product of the viscosity of the fluid and the velocity gradient, which can be simplified as:

$$S = \eta \times \frac{dv}{dx} \approx \eta \times \frac{v}{L}$$

The viscous force $(F_v)$ is obtained by multiplying the viscous stress $S$ with the surface area $A$.

$$F_V = A \times S \approx L^2 \times \eta \times \frac{v}{L} = \eta v L$$

Reynolds number $R$ is thus defined as follows:

$$R = \frac{F_I}{F_V} \approx \frac{Dv^2 L^2}{\eta v L} = \frac{DvL}{\eta} = \frac{Density \times Velocity \times Length}{Viscosity}$$

Swimming in a high viscosity fluid (large $\eta$) such as syrup results in a flow of low Reynolds number. On the other hand, a "short" microbe (small $L$ as for a paramecium) swimming at a slow speed (small $v$) is another format of low Reynolds number flow even in water with its low viscosity (small $\eta$). Although the conditions are totally different, the two flows are similar in terms of fluid mechanics. The paramecium may feel that water is just as viscous as a salad oil to a human being.

The swirling style of swimming of the paramecium is actually a smart skill that increases its effective length for conquering viscosity, much similar to a spinning American football. From a wind tunnel test for an aircraft, to the swimming pattern of a microbe, the concept of Reynolds number is a somewhat rough but effective analytical tool in fluid mechanics.

# Appendix E

## Thermal Compensation for Conductivity

From the viewpoint of a human being, both the length and velocity of a microbe are small, so is the Reynolds number of its flow (Appendix D).

$$R = \frac{DvL}{\eta} = \frac{Density \times Velocity \times Length}{Viscosity}$$

On the contrary, the microbe will feel a fluid such as water that is certainly not viscous to us, has a high viscosity. Now, you can think an ant swimming on water is just like a human being swimming on syrup. Similarly, an ion with charge Z migrating in water under an electric field ($E$) is exactly the case of a low Reynolds number flow. For a flow with Reynolds number smaller than 1, Stokes' law holds:

$$ZeE = 6\pi r\eta v$$

The electrostatic force on the left is counteracted by the viscous drag on the right-hand side, and the ion soon reaches its terminal speed, $v$. Where, $e$ is the unit charge of $1.6 \times 10^{-19}$ Coulomb; $r$ is called the Stokes' radius for the "hydrated" particle (ion), somewhat larger than the geometrical radius.

Viscous fluids are those with stronger intermolecular attractive forces, such as the hydrogen bonds in glue. To "swim" in a fluid, you must first tear apart the molecules by breaking the intermolecular bonds, which is the origin of the activation energy $E_a$ for viscosity:

$$\eta = \eta_o e^{\frac{E_a}{RT}}$$

From the above equation, the higher the temperature, the lower the viscosity, and the mobility ($\mu$) of an ion can be solved for as:

$$\mu = \frac{v}{E} = \frac{Ze}{6\pi r \eta} = \frac{Ze}{6\pi r \eta_o} e^{-\frac{Ea}{RT}} = \frac{Ze}{6\pi r \eta_o} e^{\frac{RT}{Ea}}$$

Differentiating mobility against temperature, we will have:

$$\frac{d\mu}{dT} = \left( \frac{R}{E_a} \right) \frac{Ze}{6\pi r \eta_o} e^{\frac{RT}{E_a}} = \frac{R}{E_a} \mu$$

Consequently,

$$\frac{\Delta\mu}{\Delta T} = \frac{\mu_T - \mu_0}{T - T_0} = \frac{R}{E_a} \mu_0$$

Where $\mu_T$ and $\mu_0$ are the mobility at $T°C$ and the initial temperature $T_0$, respectively. Solving for $\mu_T$, we get:

$$\mu_T = \mu_0 \left( 1 + \frac{R}{E_a} \Delta T \right)$$

Consequently,

$$G_T = G_0 \left( 1 + \frac{R}{E_a} \Delta T \right)$$

Where $G_T$ and $G_0$ are the conductivity at $T°C$ and the initial temperature $T_0$, respectively.

The value of $R/E_a$ for water is about $0.02°K^{-1}$, so the conductivity will increase about 2% per degree of temperature increase. The 2% is called the thermal compensation coefficient for conductivity which can be obtained experimentally.

# Debye-Hückel Theory

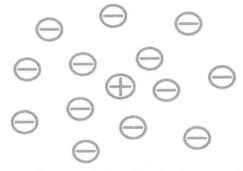

**Center Ion and the Ionic Atmosphere
of its Counter Ions**

Taking a close look with a nano-meter scale at the center ion and its surrounding ionic atmosphere, at a small distance of $r$ away from the center ion there will be a deviation from electroneutrality with the charge density ($\rho$) defined as the following:

$$\rho(r) = F \sum_{Cations} Z_i C_i(r) - F \sum_{Anions} |Z_i| C_i(r) \neq 0$$

Counter ions in the ionic atmosphere are attracted by the center ion; as a consequence, both the local potential ($\varphi$) and charge density ($\rho$) will drop more quickly in an environment of higher ionic strength ($\mu$). This tendency is counteracted by the Brownian motion of the ions and finally reaches a balance between the electrostatic attraction and thermal motion.

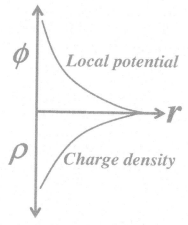

**Change in Local Potential and Charge Density
with the Distance (*r*) from the Center Ion**

Notice that there are the discontinuities of both functions at the position of the center ion ($r = 0$); the local potential will be infinity and the charge density will suddenly shift from negative to positive.

Based on Boltzmann's distribution law, the ratio of the change in charge density ($\Delta\rho$) to the change in local potential ($\Delta\varphi$) along the distance *r* can be approximated as:

$$\frac{\Delta\rho(r)}{\Delta\phi(r)} \cong -\frac{2F^2}{RT}\mu$$

For ionic strength lower than 1 mM, the approximate local potential calculated from the above relationship will decay as a function of *r*:

$$\phi(r) = \frac{\alpha}{r}e^{-\frac{r}{\beta}}$$

Where $\alpha$ is constant; $\beta$ is called the Debye radius or Debye length that is related to the thickness of the ionic atmosphere. The Debye radius $\beta$ is given by:

$$\beta = \sqrt{\frac{RT\varepsilon}{2F^2\mu}}$$

111

Substituting the permittivity of water $\varepsilon$ into the above equation, we will have a Debye radius at 25°C, 1mM of about 0.3nm.

The decay of charge density is given by:

$$\rho(r) = \left( -\frac{2F^2}{RT}\mu \right)\frac{\alpha}{r}e^{-\frac{r}{\beta}} = \left( -\frac{\varepsilon}{\beta^2} \right)\frac{\alpha}{r}e^{-\frac{r}{\beta}}$$

The center ion is "nestled" in the "atmosphere" of its counter ions, so some of the "activity" is lost. The free energy should be modified to be the following:

$$G = G^O + RT\ln\gamma C = G^O + RT\ln a$$

Where $a$ is the activity of the chemical species which is related to its concentration $C$ by a constant $\gamma$, the activity coefficient.

For solutions of low ionic strength, $\gamma$ is approximately 1, and the activity is practically equivalent to the concentration. For solutions with the ionic strength in the order of mM, $\gamma$ for an ion with charge $Z_i$ is given by:

$$\gamma = e^{\frac{-Z_i^2 F^3}{2\pi N_A}\sqrt{\frac{\mu}{(2RT\varepsilon)^3}}} \leq 1$$

$N_A$ is the Avogadro's constant. Substituting all the constants at 25°C, we get:

$$\gamma = e^{-Z_i^2\sqrt{\frac{\mu}{0.727M}}}$$

For a binary solution, the activity is usually expressed as the mean ionic activity, $a_\pm$.

$$a_\pm = (a_-)^{\frac{Z_+}{Z_+ + |Z_-|}}(a_+)^{\frac{Z_-}{Z_+ + |Z_-|}}$$

The same is used for the activity coefficient.

$$\gamma_\pm = (\gamma_-)^{\frac{Z_+}{Z_+ + |Z_-|}} (\gamma_+)^{\frac{Z_-}{Z_+ + |Z_-|}}$$

For example, the solution of $Al_2(SO_4)_3$ is composed of $Al^{3+}$ and $SO_4^{2-}$ with the mean activity coefficient of:

$$\gamma_\pm = (\gamma_-)^{\frac{3}{3+2}} (\gamma_+)^{\frac{2}{3+2}}$$

## Microelectrodes & their Fabrication

With one of the dimensions of the electrode surface being of micro-meter order, the diffusion to a microelectrode is hemi-spherical rather than the planar diffusion toward a macro-electrode.

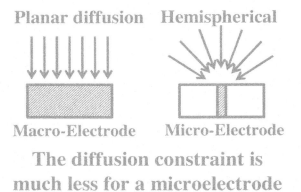

Planar diffusion    Hemispherical

Macro-Electrode    Micro-Electrode

The diffusion constraint is much less for a microelectrode

For hemi-spherical diffusion, the redox species is supplied quickly and an electrode process can be performed without a diffusion constraint. The characteristic current decay after the peak in the CVgram will not appear even with a high scan rate.

The current signal obtained with a microelectrode may be small, but the current density is extremely high. The current does not contain the unnecessary non-Faradaic component and the signal decay from the diffusion constraint. Therefore, a microelectrode is a very powerful tool in investigating Faradaic processes. The

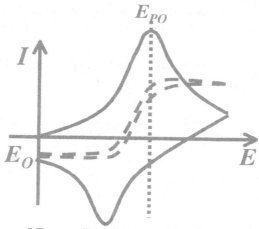

**CVgrams of Reversible System Obtained with Macro (Solid Curve) and Microelectrode (Dash curve)**

following is a standard micro-fabrication procedure for a planar microelectrode.

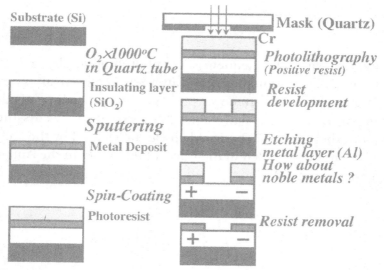

# Standard Procedure for Fabricating a Microelectrode Pair

115

The major drawback in the above procedure resides in the etching process. Since the electrode for electrochemical usages is generally made of noble metals such as platinum, noble metals are not readily etched with ordinary etching solutions. Therefore, the following lift-off process was considered:

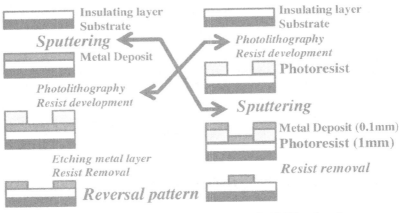

## Standard Procedure *vs.* Lift-Off Technique with the Same Mask

By reversing the sequence of photolithography and then sputtering, the unwanted metal pattern is removed at the same time during the resist removal process. The procedure needs no etching process, but a flaw called "step coverage" occurs during the sputtering process. The problem can be cleared by the "reverse taper" technique detailed in the relevant textbooks in the Appendix "Suggested Textbooks."

## Dielectrophoresis

Due to the conservation of electroneutrality, the charge/mass ratios of particles tend to decrease as the particles get larger. As a consequence, D. C. electrophoresis works well only for particles up to nano-meter scale, such as proteins and nucleic acids. For micro-scaled dielectric particles such as bacteria and cells, "dielectrophoresis" using an A. C. field becomes the choice. A typical schematic diagram is as follows:

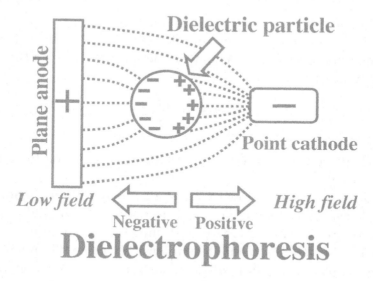

Since the electrophoretic process is not based on the absolute strength of the electric field but the gradient, an electrode pair with

unequal surface areas is used. An unbiased A. C. voltage is imposed on the electrodes, which alternatively induces a temporary dipole moment on the particles. As a result, some particles migrate toward the high field (high in the density of field lines), and the migration is defined as positive dielectrophoresis. Of course, the opposite migration toward the low field is called negative dielectrophoresis. The direction and strength of the migration force ($F$) is given by Pohl's equation:

$$F = 2\pi r^3 \varepsilon_m \, \mathrm{Re}[K(\omega)] \nabla E^2$$

Where $r$ is the radius of the particle; $\varepsilon_m$ is the permittivity of the medium; $\nabla E$ is the gradient of the root mean square of the field strength; $\mathrm{Re}[K(\omega)]$ is the real part of the Clausius-Mossotti factor that is given by:

$$K(\omega) = \frac{\varepsilon_p{}^* - \varepsilon_m{}^*}{\varepsilon_p{}^* + 2\varepsilon_m{}^*}$$

Where $\varepsilon_p{}^*$ and $\varepsilon_m{}^*$ are the complex permittivities of the particle and medium, respectively. A complex permittivity ($\varepsilon^*$) is a function of the frequency ($\omega$):

$$\varepsilon^* = \varepsilon - \frac{j\sigma}{\omega}$$

Where $\varepsilon$ is the conventional permittivity; $\sigma$ is the conductivity of the medium.

Since the Clausius-Mossotti factor is a function of frequency, it is possible to "tune" the direction and strength of the migration force by the frequency of the field for separating different particles.

With a suitable "separation frequency" in place, the remaining task is to create a higher gradient to enhance the separation. Since the geometrical design of the dielectrophoresis system is sometimes restricted, one may consider elevating the voltage to achieve a higher field and gradient. However, unwanted redox processes may occur in a voltage ($V$) higher than tens of mV. A solution for this dilemma is to use a micro-fabricated system to minimize the $L$ (length) for a larger $E$ (field strength) as in the following relationship:

$$E = \frac{V}{L}$$

Therefore, it will be much easier to use a micro-fabricated electrode and a microsystem for observing a dielectrophoresis phenomenon.

# Electrodialysis & Ion Exchange Membranes

As stated in the chapter on membrane potential, a difference in concentration may result in a difference in potential across the membrane, and that underlies the principle of the concentration cell. Is it possible to reverse the process by imposing a voltage to create a concentration difference? Electrodialysis gives a concrete answer.

It is said that table salt in Taiwan is made by electrolysis of sea water. The truth is a totally different story. The filtered sea water is first introduced into the following electrodialysis chamber:

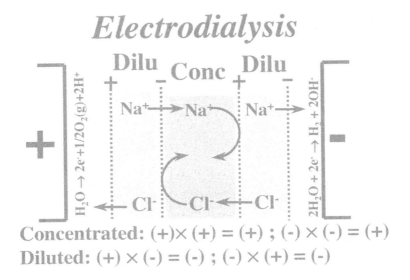

*Electrodialysis*

Concentrated: $(+) \times (+) = (+)$ ; $(-) \times (-) = (+)$
Diluted: $(+) \times (-) = (-)$ ; $(-) \times (+) = (-)$

Between the titanium electrodes, cation selective membranes (indicated as "-") and anion selective membranes (indicated as "+") are alternatively inserted. Cations migrate toward the negative electrode but can only pass through cation selective membrane, so the cations will be trapped in the "concentrating" compartments. A similar story happens to anions that will be also trapped in the same compartments. The concentrated sea water in the concentrating compartment is then drawn and heated to obtain table salt by evaporation.

Your curiosity may next be drawn to wonder why an anion selective membrane allows only the passage of anions.

### *Semi-Permeable*
### *Anion Exchange Membrane*

The surface of an anion selective membrane possesses immobilized positive charges such as alkyl ammonium ions ($R-NH_3^+$) and pores just the size of the hydrated anion such as chloride, $Cl^-$ $\bullet nH_2O$. Within the channel inside a pore on a chloride-selective membrane, there must be an array of chloride from inside to the outside of the membrane. Once a chloride ion outside the membrane hits the first chloride ion of the array, the "momentum" will be transferred to the last one and finally become a "chloride input" to the internal compartment, just like the cue ball hitting an array of billiard balls.

# Fuel Cell: Burning without Fire

Combustion is a violent oxidation phenomenon accompanied by the emission of heat; it requires fuels, oxidants and an instant high temperature over the flash point, just like the striking of a matchstick. The earth is abundant in the handy oxidant, the oxygen molecule, to oxidize the carbon-containing fuels such as wood, coal and petroleum. Burning seems to be the most convenient way to obtain energy, but the process can hardly be controlled for a sustained energy output or to prevent explosion.

If we can harness the violent oxidation process in an electric mode, then we can turn "match" into "switch" for the trigger and control the complicated heat transfer problem with a single resistor. Fuel cells make the dream come true.

It is certainly possible for fuel cells to use the aforementioned carbon-containing fuels, but that will emit the eco-unfriendly greenhouse gas.

$$C + O_{2(g)} \rightarrow CO_{2(g)} (\text{Greenhouse gas})$$
$$2H_{2(g)} + O_{2(g)} \rightarrow 2H_2O_{(1)}$$

Therefore, hydrogen gas is the ideal fuel for environmental concerns. The remaining problem is to retrieve the energy from the following half reactions:

$$2H_{2(g)} \rightarrow 4H^+{}_{(aq)} + 4e^-{}_{(S)} (\text{Oxidation, Anode, Negative})$$
$$O_{2(g)} + 4H^+{}_{(aq)} + 4e^-{}_{(S)} \rightarrow 2H_2O_{(1)} (\text{Reduction, Cathode, Positive})$$

The difficulty in designing a system for the above electrochemical process is to provide an interface on which the reactants in gas phase ($O_2$ and $H_2$), liquid phase ($H^+$ and $H_2O$) and solid phase (the electrons) will mix and react.

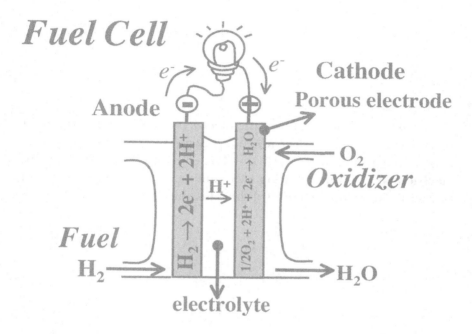

In the above system, a pair of porous electrodes serves as the three-phase reactor for the coupled redox reaction. To reduce the volume and internal resistance, the electrolyte between the electrodes can be replaced with a cation exchange membrane used as a spacer, such as the Nafion™ membrane of DuPont Co.

## Ohmic Heating

Actually it is the heat, not the electricity that causes permanent tissue damage after an electric shock. Caught by an electric mosquito racket, the pitiful insect will smell like it was "charred" with an electric spark. In the following experiment, an innocent shrimp will be sacrificed for the electric (ohmic) heating method:

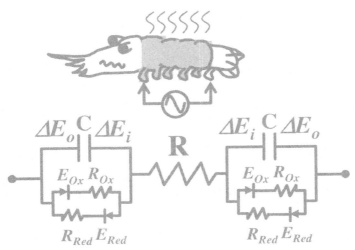

## Schematic Diagram and Equivalent Circuit of Ohmic Heating

After inserting stainless nails both into the head and tail of a shrimp, the nails are connected to a domestic A.C. voltage source (110 V, 60 Hz) through a switch for safety. The raw (greenish) tissue between the nails will turn red in the seconds immediately after switching the power on, but the remaining parts are still "innocent." The "guilty" tissue was actually heated by passing an A.C. current as illustrated in the above equivalent circuit.

As already learned in the chapter on interface impedance, there is a non-Faradaic charging current (the upper pathway in the figure) and a Faradaic redox current (the lower one) flowing through the electrodes. To simplify the explanation, a pulsatile A.C. voltage is applied instead of the usual sinusoidal voltage.

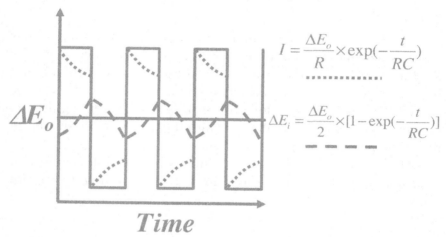

**Oscillating Potential Steps ($\Delta E_o$) and the Resulting Charging Current ($I$) and Interfacial Potential ($\Delta E_i$)**

With a large electrode (high RC, the time constant) and a high frequency (small $t$), the A.C. charging current (the upper pathway) will not rapidly decay before the next potential step.

$$I = \frac{\Delta E_o}{R} \times \exp\left(-\frac{t}{RC}\right)$$

Similarly, the interfacial potential will not extensively rise before the next voltage reversal.

$$\Delta E_i = \frac{\Delta E_o}{2} \times \left[1 - \exp\left(-\frac{t}{RC}\right)\right]$$

If the interfacial potential does not rise to the extent with enough overpotential, significant redox reactions (the lower pathway) will not occur. In other words, the electrode will not be sufficiently polarized to conduct a redox reaction. Therefore, for high frequency ohmic heating, no redox reaction is supposed to occur and the Warburg element is neglected in the equivalent circuit.

The power of heating can be calculated as follows:

$$P_{heat} = I^2 R = \frac{\Delta E_o^2}{R} \times \exp\left(-\frac{2t}{RC}\right)$$

This heating method is sometimes demonstrated in a magic show and has become an important technique in food processing. The food itself will act as the resistor (R) in the equivalent circuit, and heat transfer will not be a problem even if a viscous material is to be heated. With negligible redox process, no electrode fouling or food contamination problems will occur with the power at a frequency in the range of kHz.

The electric chair used for executions is a horrible application of ohmic heating which also uses an A.C. power source to prevent the fouling of the cables.

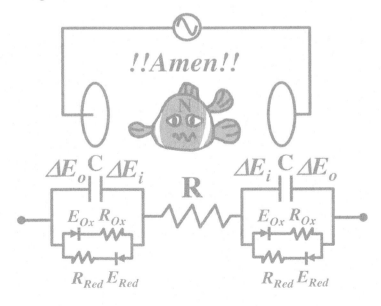

# Appendix L

## Suggested Textbooks

- **Bard & Faulkner** (1980) *Electrochemical Methods: Fundamentals and Applications.* John Wiley & Sons.
- **Barrow** (1988) *Physical Chemistry.* McGraw-Hill.
- **Brett & Brett** (1996) *Electrochemistry: Principles, Methods and Applications.* Oxford.
- **Koryta, Dvorak & Kavan** (1993) *Principles of Electrochemistry.* John Wiley & Sons.
- **Lambrechts & Sansen** (1992) *Biosensor: Microelectrochemical Devices,* Institute of Physics Publishing.
- **Oldham & Myland** (1994) *Fundamentals of Electrochemical Science.* Academic Press, Inc.
- **Rashid** (1999) *Microelectronic Circuits: Analysis and Design.* PWS Publishing Co.

# Appendix M

# Preface of The Original Japanese Version

　　近年、バイオテックの急速的発展に連れて、電気化学の手法は理解していないままに使われてきた。pH電極を始めとしてのイオン選択性電極、溶存酸素電極、バイオセンサー、電気泳動ないし遺伝子解析によく使われる毛細管電気泳動、バイオチップなどの技術は全て電気化学の理論に基づいている。電池技術の発展はさらに電気化学の重要性を高めてきた。

　　しかし、物理化学の一分野の電気化学を紹介する教科書は、物理化学の体質である厳密性を守るため、物理量等をうるさいほど定義するので、かえて入門の支障になる。たとえば、濃度を取り扱うには、物理化学では「活量」を使用する。そのためにデバイ―ヒュッケル(Debye-Hückel)の理論などを導入しなければならない。すると、話は話題からずれて長くなるにも関わらず、気の短い近頃の学生さんの神経を殺すことになる。

　　数学は科学者の共通な言葉であり、わざと数学的取り扱いを避けるよりも「神経質でない」電気化学を一息で紹介した方が効率的であると筆者が思う。そこで、本書は理論、実験、応用よりも電気化学のエッセンスを紹介することを目的とした。内容は、バイオテック関連分野によく使われる電気化学手法を中心にし、できるだけ短くて肝要で電気化学をイメージさせるものである。

　　具体的には、膜電位、伝導率、酸化還元電位、電気化学反応と電気泳動などをセンタードクトリンとして紹介しながら、pH電極、溶存酸素電極、毛細管電気泳動などの技術の原理を明らかにする。話を滑らかにするため、それに必要となる背景知識と応用を後回して付録で紹介する。

　　2003年6月

陳　力騏

# Preface in Chinese

前言

　　從最簡單的pH電極到基因體計畫所用的毛細管電泳技術，你可能沒有注意到幾乎所有生物相關技術都是基於電化學原理。而最近電池科技的發展更提高了這個知識領域的重要性。

　　但是，對於沒有堅實化學背景的學生而言，想要了解電化學背後的原理卻是一條漫漫長路。由於是物理化學的一條分枝，一般電化學教科書對於所牽涉的物理量都存在物理化學特有的嚴謹度、甚至有點神經質的定義。例如，為了更精密地計算自由能，則使用「活性」來代替「濃度」，如此一來就必須先介紹與本文較不相關的Debye-Hückel理論，這樣勢必影響教學的流暢性。

　　數學是科學家共通的語言，與其刻意地閃避數學式，毋寧以較不神經質、更有效的方式作觀念的連結，以闡述特別是生醫領域中常用電化學技術的原理。作者通常於每一章後，至少設計一實驗以闡明一些可能被隱藏著的觀念，但基於篇幅，將之置於系上或研究室的網站中以供讀者下載(連同本書原始的日文版本)。此外，為了使本書更加完整，作者尚於附錄提供一些背景知識與其他應用。

　　本書最初於2003年以日文寫成，以輔助系上機械背景學生的學習。然作者於2005年在藥學系代課電分析化學時，發現Bio背景的學生對部份數學處理、電學與流體的概念仍需要強化才能適應原教材。所以，在2006年增添此部份的內容於第二版日文書中。試用兩年後，再於2008年譯為英文，並修定、補正部分內容。

　　最後，在此對於那些提供寶貴意見的系上同仁以及學生致意，並同時向各位讀者為忍受本人的台灣風英文及拙劣的插圖致歉。

陳力騏

2008年2月28日

生物感測研究群
生物產業機電工程學系
台灣大學

# Index